THICK FILM TECHNOLOGY AND CHIP JOINING

PROCESSES AND MATERIALS

IN ELECTRONICS

A series of monographs, texts, and reference books covering areas and topics of current interest in electronics

Edited by HOWARD H. MANKO

Volume I
THICK FILMS TECHNOLOGY AND CHIP JOINING

LEWIS F. MILLER

OTHER VOLUMES IN PREPARATION

CLEANERS AND CLEANING IN ELECTRONICS
HOWARD H. MANKO

PLATING IN THE ELECTRONIC INDUSTRY
JAMES LANGAN

HYBRID MICROELECTRONICS
PROCESSES AND MANUFACTURING TECHNIQUES
ALFRED LEVY

CERAMICS IN ELECTRONICS
EARNEST KASTENBEIN AND
FRANK PIRIGYI

THICK FILM TECHNOLOGY AND CHIP JOINING

LEWIS F. MILLER
International Business Machines Corporation

GORDON AND BREACH, SCIENCE PUBLISHERS, INC.
New York   London   Paris

Copyright © 1972 GORDON AND BREACH, Science Publishers, Inc.
440 Park Avenue South, New York, N.Y. 10016

Library of Congress Catalog Card Number: 79-175344

Editorial office for Great Britain
  Gordon and Breach, Science Publishers Ltd.
  12 Bloomsbury Way
  London W.C. 1 England

Editorial office for France
  Gordon and Breach
  7-9 rue Emile Dubois
  Paris 14ᵉ, France

ISBN 0 677 03440 7 (cloth): All rights reserved. No part of this book may be reproduced or utilized in any form or by any means, electronic or mechanical, including photocopying, recording, or by any information storage and retrieval system, without permission from the publishers.

Printed in the United States of America

Table of Contents

	Introduction	1
Chapter 1	Screening and Paste Transfer ...Factors controlling the printed result	5
Chapter 2	Thin Film Conductors ...A brief survey	25
Chapter 3	Silver Palladium Electrodes ...A detailed study of formula and properties	33
Chapter 4	Ternary Alloy Electrodes ...The effects of formula changes on properties with emphasis on three types: (1) Au:Pt, (2) Au:Pt:Pd, (3) Ag:Au:Pd.	61
Chapter 5	Glaze Resistors ...The influence of the vehicle and paste preparation on resistor properties	81
Chapter 6	Controlled Collapse Reflow Chip Joining ...An intensive look at the materials and process of a specific chip joining method	101
Chapter 7	Survey of Chip Joining Technique ...An overview of many methods	127
Chapter 8	Critique of Chip Joining Techniques ...A critical comparison and evaluation	155
Chapter 9	Powder Interconnections ...Connecting layers in multilevel ceramic substrates	183
	Bibliography	205
	Index	215

INTRODUCTION

The complex field of electronics and semiconductor packaging has proliferated many technologies – some of which have seen commercial success – and others which have not, but which are still of considerable interest to provide guidance for future development. Thick films are of the former variety: highly successful both in technology and in commerce, with wide usage throughout the industry. To identify thick films more clearly, let us define them as those passive components which are applied to a substrate from a fluid, and which are dried, cured, or fired to achieve their final properties. This delineates thick films as deriving from a liquid-to-solid transformation, which does not involve plating, vacuum deposition, evaporation methods, sputtering, or other processes – despite the thickness of the deposit.

The most widely used process for depositing the paste-like thick film materials is screening (or doctor blading) and such passive components as conductors, resistors, dielectrics, and capacitors have been developed.

Except in special circumstances, circuits are not composed only of such passive components, but also utilize active devices such as transistors, diodes, or integrated circuits. The substrates on which all of these functioning parts are mounted must also be interconnected to other circuits and finally comprise parts of machines. Therefore, the full understanding of thick films suggests an insight into other aspects of electronic packaging, chip joining, and module fabrication.

This book describes some detailed aspects of thick film materials as well as the processes that result from their usage. The first four chapters (section one) deal predominately with the materials and how they function, while the next four chpaters (section two) elaborate on module fabrication processes.

This book is not an encyclopedic description of thick film technology; it represents one line of experiments which have shown to be viable, both from the commercial and the technological standpoint. All of the work described here (except for the surveys) was performed by the author and his associates at the IBM Laboratory in Poughkeepsie and East Fishkill, New York. Large numbers of circuits have been manufactured (not only by IBM, but also by others in the electronics industry) with such results of these development programs as silver/palladium electrodes (Chapter 3), and controlled collapse reflow chip joining (Chapter 6).

The first chapter describes some aspects of the screening process itself which influence the properties of the components. It shows why unexpected changes in resistance are sometimes observed, and why there are certain limits to the screening process. The remainder of the first section is an intensive study of the raw materials and technology of the pastes themselves, and is mainly confined to air firing systems which dictate the use of non-oxidizing noble metal conductors. Four types of conductors are discussed in considerable detail: Silver:palladium, gold:platinum, gold:platinum:palladium, and silver:gold:palladium. The influence of the raw material properties on the functionality of the conductors or resistors can be observed in many - sometimes unexpected - ways. Indeed, the second section of Chapter 5 also shows how paste formulation must be modified for special module fabrication procedures, and should be considered part of the materials exposition of this book. Several examples of interactions between conductors and resistors are given in Chapters 3 and 4.

In section two, Chapter 6 describes the details of a solder reflow chip joining process which is of particular interest to semiconductor memory manufacturers. In order to put this particular technique into a proper prospective, Chapter 7 surveys many of the other chip joining methods which have been described in the literature. Considerable detail about module processing is included in this, and the following chapter - which critically evaluates the various chip joining procedures. These two chapters discuss several technologies relating to the fabrication of the modules themselves, including non-thick-film means to apply circuitry. Since ultimately rather complex modules may be required to interconnect many chips and sophisticated electrical functions, the last chapter considers multilayer modules and how a particular type of multilayer ceramic module can be interconnected at low cost.

The author is grateful for the contributions of his co-workers to the technology described here, and he recognizes that it is unfeasible to thank them all individually in a short space. However, he would particularly like to express his appreciation to the following individuals who directly aided in the developments described here: Mr. G. J. De Paolo, Mr. S. A. Milkovich, Mr. K. N. Neisser and Dr. A. H. Mones; also Mr. A. A. Valachovic for editing the original manuscripts and his valuable assistance in making the book a reality.

                                                            Lewis F. Miller
                                                            31 March 1971

Chapter 1

SCREENING AND PASTE TRANSFER

This chapter describes some simple experiments intended to clarify aspects of the screening process. It addresses the relationships of the amount of paste deposited on a substrate by screening, to certain aspects of the screening process - - such as the mesh size, paste rheology, line width, etc. It is not intended to be a vigorous mathematical or mechanical analysis of the process. The conclusions can be extrapolated not only to an understanding of a part of the mechanism of paste transfer, but to a prediction of screened thickness - - and also to methods for increasing the uniformity and thickness of screened images.

A simple experimental procedure was used to examine the transfer process. Patterns were screened by a hand-held squeegee through various meshed screens and masks with several rheologically different inks, in formats which include different width lines and squares. The resulting images were dried at 90°C for one half hour and the thicknesses were measured with a Zeiss light section microscope.

Three patterns were used:
1. Pattern A contained a variety of line widths; the nominal 5, 10, 15 line widths were measured, as well as a 50 mil square (Fig. 1).
2. Pattern B contained a group of different width rectangles; the 10 x 20, 20 x 30, 30 x 40, 40 x 120 and 100 x 120 mil rectangles were measured across the center of the smaller dimension (Fig. 2).
3. Pattern C contained two columns of lines from 5 to 45 mils wide, in 5 mil increments, with 10 mil spaces between lines. The wide and narrow lines were alternated and one column was the opposite of the other (Fig. 3).

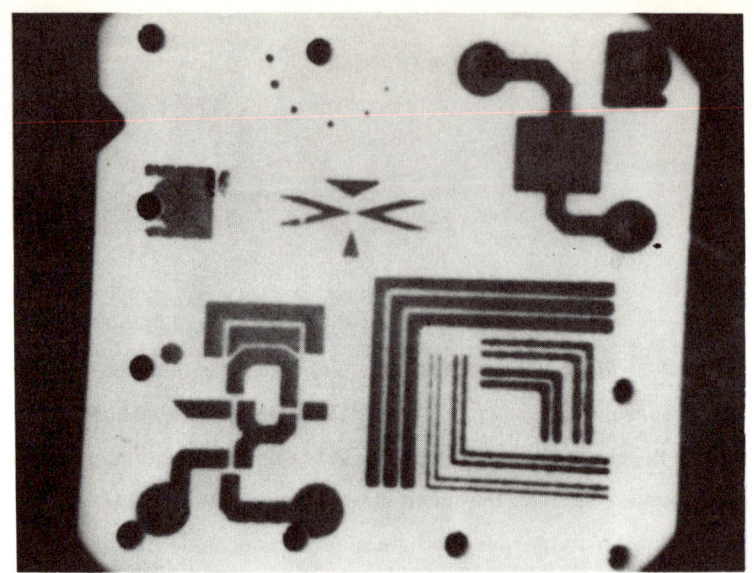

Fig. 1.   Screened pattern A.

Fig. 2.   Open mask, pattern B.

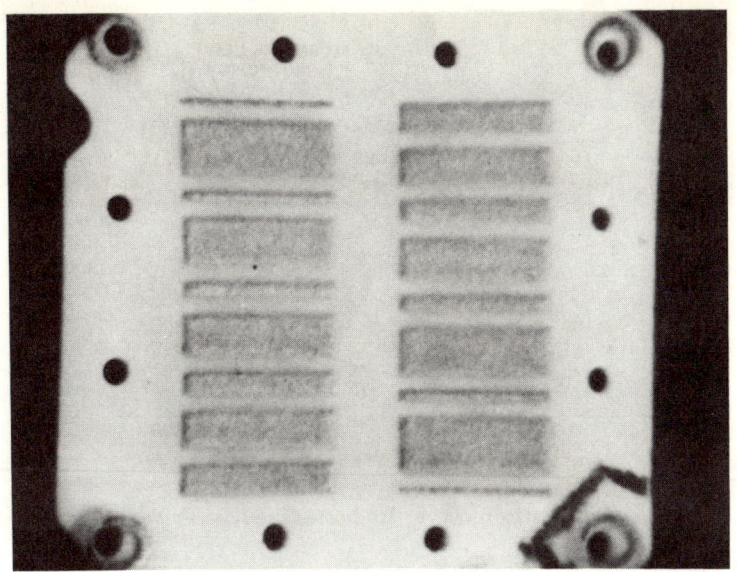

Fig. 3. Fired test pattern C.

Several types of paste were used with different rheological properties: A is a paste with relatively polar vehicle, a wetting agent, and a flow stopper, which produces profile type A of Fig. 4; B is a paste with a nonpolar vehicle, considerable flow and Newtonian behavior (that is, the stress-strain curve is linear over a broad range) with a yield value (Bingham body) - - which produces profile type B of Fig. 4; C is a paste which is less Newtonian, but which also produces profile B; D is a gelled paste with strong pseudoplasticity (a marked decrease in strain with an increasing rate of shear), which produces profile type D of Fig. 4. The generic rheological curves of these inks are shown in Fig. 5; thixotropic hysteresis is omitted.

Fig. 4. Profiles of screened images.

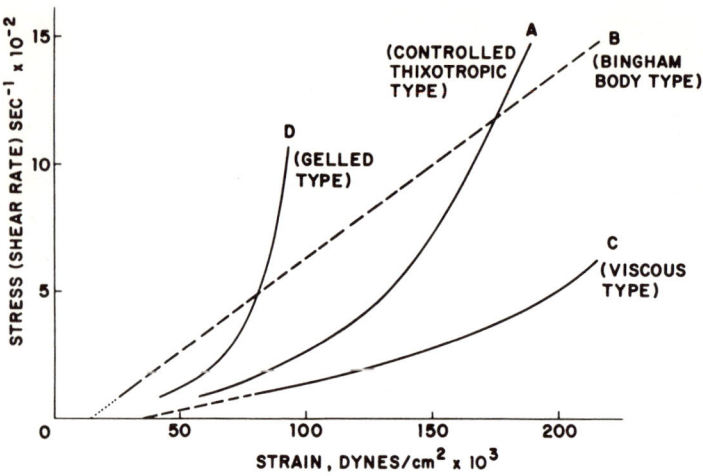

Fig. 5. Stress-strain curves for pastes used.

The data from these experiments are typified in Table 1. They indicate that a complex relationship exists between the line width and line thickness. This relationship is affected (to a greater or lesser degree) by many factors in the screening process, which will be discussed in sequence. The typical curve for mesh screens can be represented by Fig. 6, which shows three distinct sections (a, b, c) and two transition points (1, 2).

Section a - Section a is apparently linear: line thicknesses increase directly with line width. For conductor patterns, this section is the important one, extending to about 15-20 mil wide lands. Its slope varied with the screen mesh (Fig. 7), apparently increasing with mesh number. Thus, less difference in land thicknesses occur for different width lines in this section with 325 and 400 mesh than, for example, with 105 mesh, (Fig. 8). However, this is accomplished at the sacrifice of the total transfer, which is obviously lower with the 325 or 400 mesh screens. Of course, the amount of paste deposited can be increased with greater emulsion (or cavity) thickness, but this improvement is not generally as great as significantly reducing the mesh number. A few experiments by other investigators provide the data for this conclusion.

Table 1 - Typical Thickness Data–Pattern C

Line Thickness, microns

Line Width, mils	Left Columns				Right Columns			
	1	2	3	4	1	2	3	4
5	13	12	13	13	21	19	22	18
10	21	18	16	20	24	23[a]	23	21
15	25[a]	21	24[a]	22[a]	24[a]	21	23[a]	24[a]
20	20	21[a]	20	21	22	20	21	21
25	18	18	21	20	22	21	18	21
30	19	16	19	20	19	18	21	20
35	21	17	19	20	18	14	19	16
40	18	17	19	18	17	14	20	18
45	18	15	18	16	20	15	15	18

NOTE: The heavier deposits in the right columns are attributed to squeegee skew.

[a] Thickest deposit in column.

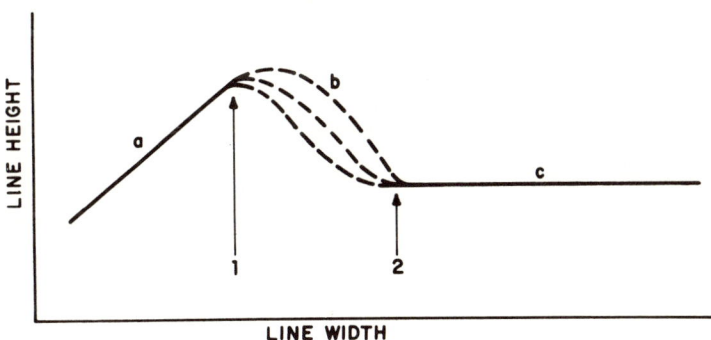

Fig. 6. The transfer curve.

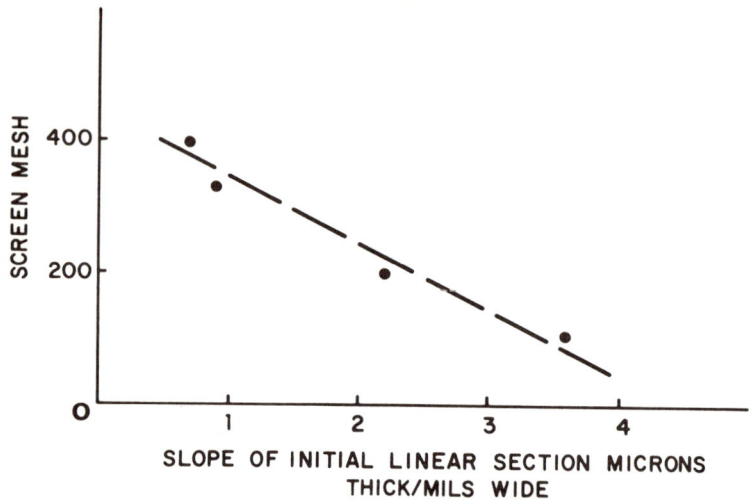

Fig. 7. "Controlled flow" paste A through various mesh screens–pattern A of Fig. 1.

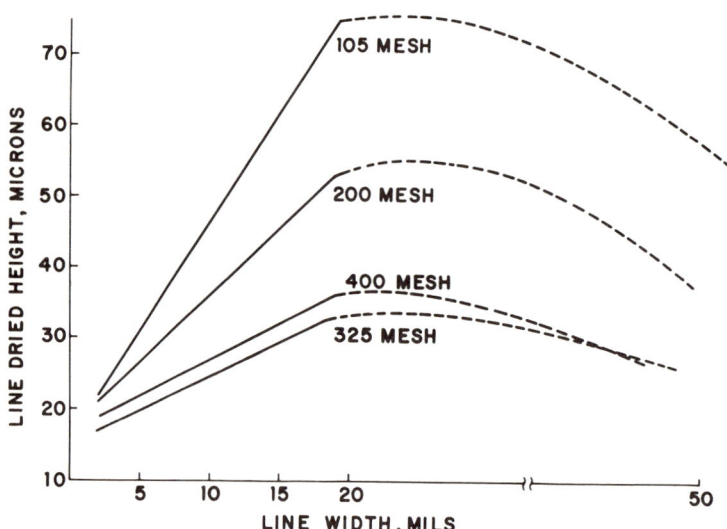

Fig. 8. Relationship of screen mesh to scope of initial linear section of Fig. 7.

It was found that 105 mesh screens produced about twice the thickness of 200 mesh screens, while a rather large increase in emulsion thickness (from 1 mil to 1.6 mils) in a 200 mesh screen produced a 40-50% thickness increase. Moreover, a 2.3 mil emulsion produced about 20% less thickness than the 1.6 mil emulsion - - perhaps due to paste immobilization (which will be discussed later). Another experiment with a Au:Pt conductor paste showed an increase in dried land height (nominal 15 mil width) from 17 microns to only 27 microns as the emulsion thickness was more than doubled from 0.7 to 1.5 mils.

E. Schneider and A. Sarab experimented with a commercial resistor paste (profile type B), screened with a pneumatically driven squeegee through screens with varied resistor sizes. As seen in Fig. 9, resistance increases in the order of 20% (interpretable as thickness decreases) were observed with increasing resistor size from 30 to 150 mils. Large differences were also obtained with variations in emulsion thickness and mesh size, in the expected directions. Resistance values for 105 mesh were always lower (greater thickness) than those for 200 mesh, even when the emulsion thickness was considerably lower for 105 mesh.

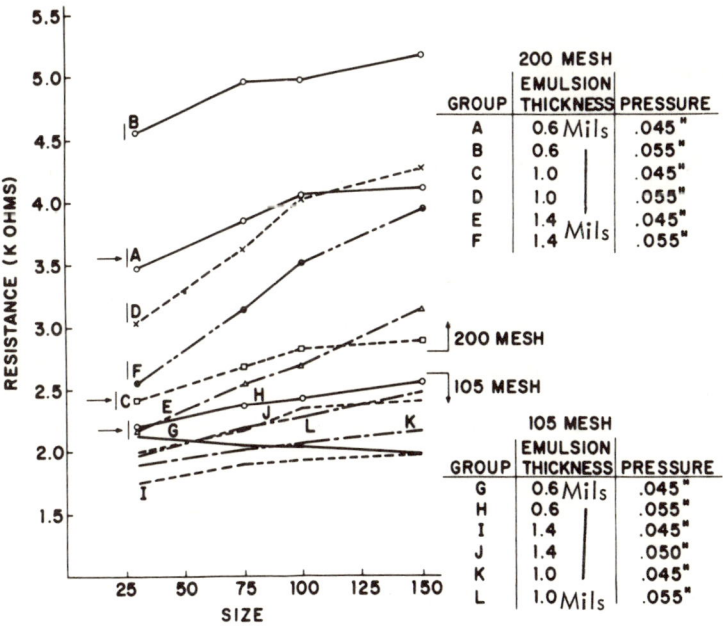

Fig. 9. Relation of resistor values with variations in squeegee pressure, emulsion thickness, and mesh size (105 and 200 mesh) (Schneider and Sarab).

Other experiments, which are too numerous to be discussed here, yielded similar results - - particularly in terms of land conductance. The effect on narrow width land conductance seems to be either a power law or two intersectioning straight lines; as lands decrease in width from 15 mils, disproportionate increases in resistance are attained from screens. A typical general relationship is shown in Fig. 10. Open masks (with no mesh), show less change in the 3-5 mil range, as shown in Fig. 11. Drastic improvements in conductance are obtained with thick open masks - - this technique has indeed been used for increasing land conductance values. Masks with critical conductance sections containing no mesh are feasible.

Section b - Section b consists of a group of possible curves which descend from the maximum at transition 1 to intersect with the relatively flat section c. The reproducibility of section b is considered to be a factor of many facets of the screening operation itself - - such as squeegee angle, pressure, speed, mesh tension, etc. Indeed this transition provides the clue to understanding why the complex transfer curve exists.

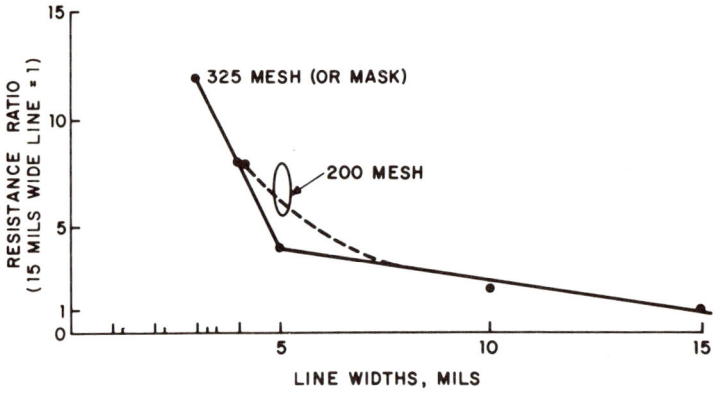

Fig. 10. Typical resistance ratios for different line widths-paste type A.

In common printing processes - - including letterpress, lithography, and intaglio - - the ink transfer from the printing plate is not complete. The split of the ink is dictated commonly by the wetting of the two surfaces (as printing plate and paper) by the ink, combined with other immobilization factors, such as surface roughness or absorption of ink vehicle into the printed surface. If both

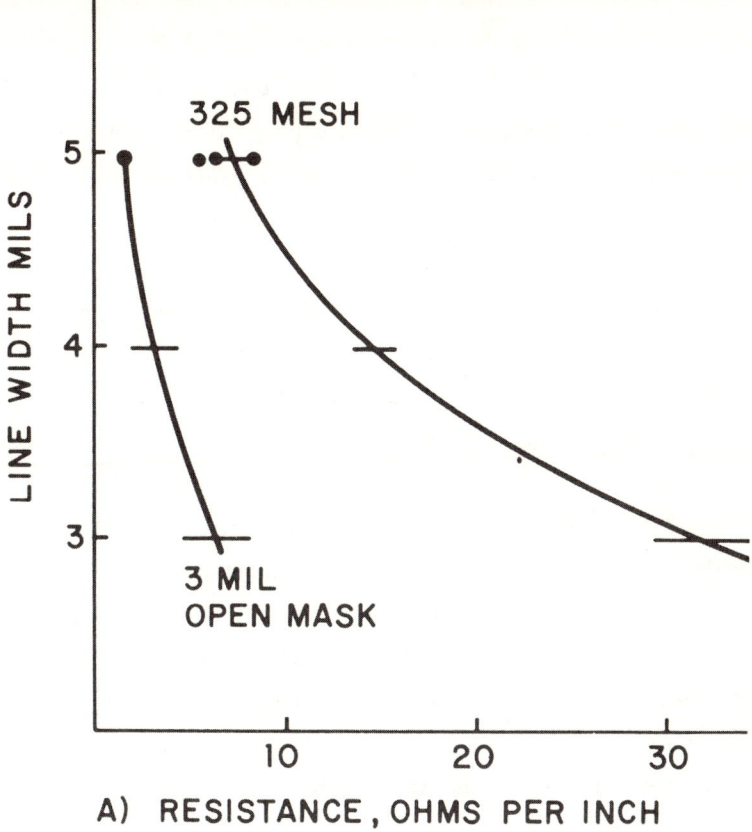

Fig. 11. Relationship of resistance to line width (range 3-5 mils)- paste type A.

surfaces are equivalent and moving at the same speed, about half the ink should adhere to either surface. Electrostatic attraction or repulsion as well as the roughness and porosity of the printed surface can be used to improve the transfe ratio.

In similar fashion, paste adheres to all surfaces it touches, (unless no wetting occurs). Some paste is immobilized on the mesh wire and on the sides of the emulsion in the cavity. The squeegee can wipe the top of the mesh fairly clean, but does not directly control the amount of paste immobilized underneath. Here the cohesive, viscous, viscoelastic, and other complex attributes of the paste rheology come into play - - combined with the speeds of the squeegee and time of the separation of screen and substrate. In gross terms, the less space between the edges of the emulsion, the greater volume of the available paste is immobilized on those surfaces. This leads to reduced transfer. As the emulsion edges are moved further apart, the greater proportional volume is transferred -- coupled with reduced capillary effects. This can account for the increased image thickness observed in section a of the transfer curve.

At transition 1 (Fig. 6), the distance of the emulsion edges may be sufficiently far apart so that two things occur: the amount of immobilized paste on the edges has become negligible compared to the total volume of paste, and the strands of the mesh begin to be depressed towards the substrate. In section b, this latter phenomenon becomes more important as the line width progresses from transition 1 to transition 2 (which is in the area of 40 mils land width). As the emulsion edges get further apart, their support at the center of the screen gets less, permitting more depression. Thus, the squeegee pressure, angle, and speed variations can form a family of curves in section b, depending on how far the mesh is depressed and the vector forces applied to the ink. Without careful control of the screening operation, section b would seem to be a difficult zone to control due to these factors. It should be recognized that the location of the transition points may vary as screen resilience and other parameters are changed. The speed of separation of screen and substrate is probably also important, since viscoelasticity of the paste is a time-dependent function.

Section C - At transition 2, the emulsion edges are essentially at "infinite" distance, where further changes in transfer are minor. Beyond this transition, in section c, thickness values do not change drastically, but in essence level out, particularly with low mesh numbers. It follows that in this region, the mesh is either depressed to the substrate (at least in the center), or a reasonably reproducible quantity of paste separates the mesh from the substrate. If the latter were true, changes in squeegee pressure should change the transfer and data has indicated that this is the case. It should be relatively simple to acquire further data to clarify this aspect.

Masks with etched mesh seem to follow a somewhat similar pattern, but with less decrease of thickness values in the progression from transition 1 to transition 2. This follows, since we would expect less mesh depression from the more rigid mask. Open masks (without mesh) show much less variation; indeed careful further scrutiny may show insignificant thickness variations once the paste immobilization characteristic of section a is overcome - - which should be at, or before, 15 mils land width.

It is evident that with very narrow lines, capillarity and ink immobilization are important factors. Depending on ink rheology, plug flow can occur with open masks with a high depth-to-width ratio. Experiments showed, for example, that for line widths up to about 10 mils, a practical limit for open mask thickness is in the order of about three to four mils with pastes similar to A. Beyond that thickness, only a portion of the paste transfers to the substrate, leaving the remainder in the mask. For example, a five mil thick open mask deposited only about 1-2 mils of wet paste A in a 4 mil wide pattern; a ten mil mask deposited about the same height of paste. However, as the line width was increased to 6.5 mils, deposits of 3 mils height were obtained. Similarly, a 20 mil thick open mask could not produce very high 20 mil wide lines - - plug flow occurred with several rheologically different electrode and glass pastes. Of course, this type of plug flow can be minimized, but then the images might flow out and be useless. The rheological compromise is thus presently a limiting factor in attaining very thick narrow lines. It appears plausible at this time to attain well defined four mil lines which are 1 mil thick after firing, by using open masks. Of course, some supports in these masks may be required for mechanical handling and can be constructed in specific locations on the mask.

Some further interesting aspects can be studied by manipulating the width-versus-thickness relationships. As seen in Fig. 12, a case can be made for a linear relationship between mesh number and the line width in section a which produced a similar thickness in section c. As the thickness of the relatively flat section c increases, the line width needed to produce a corresponding thickness in section a should also increase. The graph shows this to be true. (Table 2 shows one manufacturer's specification for steel wire used for making screens).

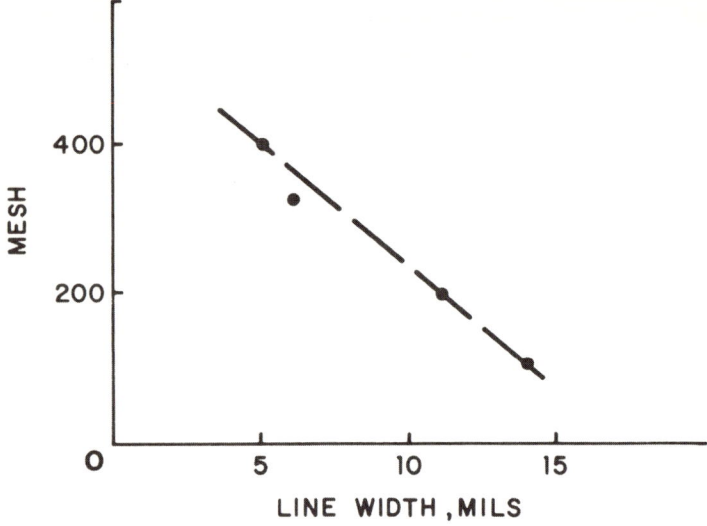

Fig. 12. Relationship of line width in initial linear sections in Fig. 7 which produced same thickness as large square (infinite width) to mesh size.

Table 2 - Specification for Steel Wire Used for Making Screens

Mesh	Wire Dia, in.	Opening, in.	Open Area, %
105	0.0030	0.0065	47
200	0.0021	0.0029	34
325	0.0012	0.0019	37
400	0.0010	0.0015	35

These relationships are graphically presented in Fig. 13. Only wire diameter and opening are nonkinked curves; the percentage of open area does not offer good predictive ability. The opening between the wires does follow the trend, permitting less capillary ink immobilization for the lower mesh numbers, but not clearly differentiating between 325 and 400 mesh. This may account for the negligible difference observed in the slope of section a in Fig. 7 for 325 and 400 mesh.

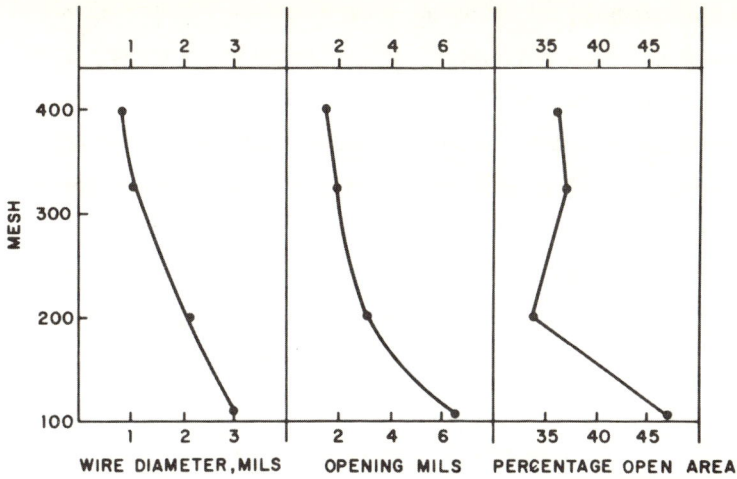

Fig. 13. Wire mesh properties.

With etched metal masks, a one mil thick mesh section is less of the total volume than with the wire screens, and less control of the transfer might be expected. No linear relationship between the slope of section a and the mesh thickness or opening could be found, but more precise experiments might determine such a relationship for the latter (Fig. 14). If true that the opening is truly the main controlling factor here, it would seem that reducing it would cause less variation in the screened deposits for different widths - - but again the same considerations apply as before: such a move creates thinner deposits and would lead to clogging difficulties. In addition, only minor opening variations in wire screens are readily available commercially. Instead, the impetus should be in the opposite direction - - eliminating the mesh entirely wherever practical.

EFFECT OF INK RHEOLOGY

Ink rheology is theoretically fascinating, experimentally frustrating, and in a practical sense, realistically only partially manipulable. Viscosity is particularly important at two rates of shear: the paste must resist being pushed between the mask or emulsion and the substrate during screening, and it should not flow beyond the original screened dimension under the force of gravity. To solve these conditions, several rheological values should be controlled:

1. At the screening rate of shear, an appropriate viscosity should be specified, which is not significantly changed by thixotropic breakdown at that shear rate.
2. A thixotropic structure should be established very rapidly (within several seconds) when active shearing is removed, to prevent running or secondary flow of the image. However, sufficient cohesion should exist to minimize plug flow, keep initial yield values low, and permit leveling of the image surface - - without allowing paste "stringing" to occur when screen and substrate are separated.

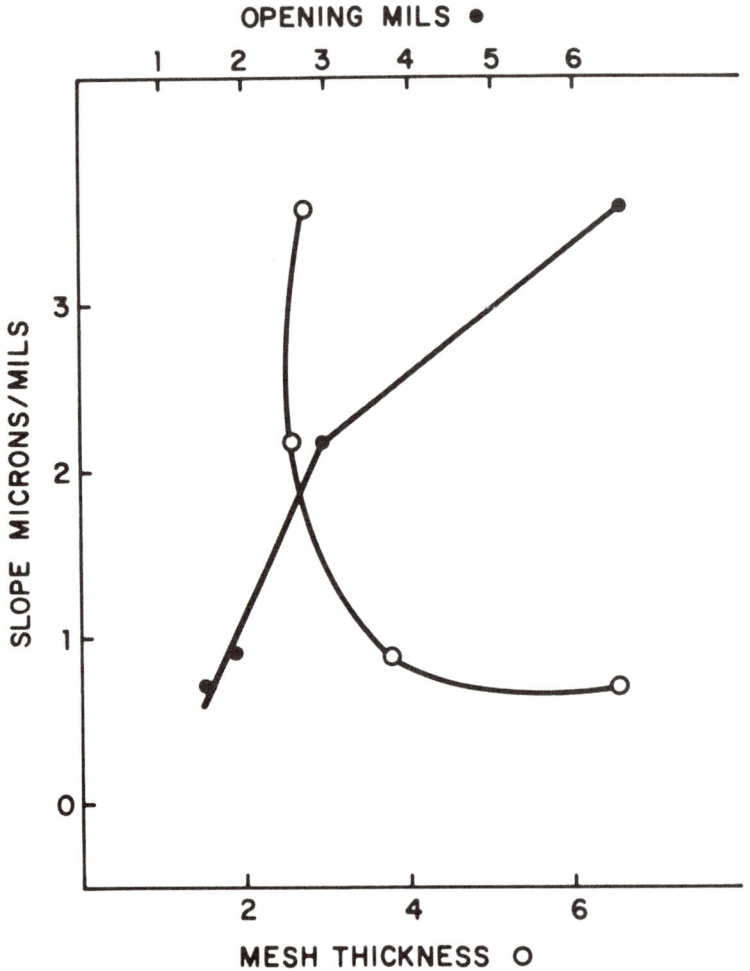

Fig. 14. Relationship between slope of section A of Fig. 9 and opening and mesh thickness.

These requirements result in paste systems that can be described as non-Newtonian pseudoplastic thixotropic bodies with yield value. Obviously, some of the pastes used in these experiments do not fit this description, and in fact do not make good screening conductive pastes, such as type B and to a lesser degree, C. Surprisingly, drastic improvements can be made in secondary flow characteristics without appreciable changes in the stress-strain curves. Additions of furoic acid and terephthalic acid for example tend to stop the secondary flow in a remarkable manner, but the rheological differences are not major and have not been adequately described. Probably the time rate of thixotropic buildup and the yield values on the descending curve are the important factors here.

A comparison of Figs. 15 and 16 shows differences in the accentuation of the transfer curve both from open masks, mesh masks, or screens when a Newtonian paste, type B, was used (actually, a Bingham body, with yield value), compared to Paste A (with furoic acid). A drastic difference was also observed with a gelled paste, type D (Fig. 17). Here plug transfer caused the entire volume of ink to be pulled from the open mask on the finer lines, while wider lines showed less transfer. I do not understand this phenomenon, but if it is repetitive such gelled structures may be very advantageous for depositing thick narrow lines through open masks. However, if mesh is used, such a paste will give clogging difficulty. Very flat resistor profiles can be obtained with such a paste and open masks (Fig. 18). In such a process, the images do not show the normal depression in the center of wide lines commonly resulting from screening mesh depression, but instead possess the sharp shoulders and extra lip caused by paste pulled up by the mask when it is lifted. This lift-up, incidentally, is probably one of the causative factors in producing thin land edges. The slope section at the image edge is relatively constant at 2-3 mils for 400-200 mesh screens. It tends to be more exaggerated with Newtonian pastes, but does not appear to be particularly related to the mesh size. Controlled flow pastes (like A) show the thin edge more than flowable pastes where the paste levels out and flows over the thin edge, hiding it.

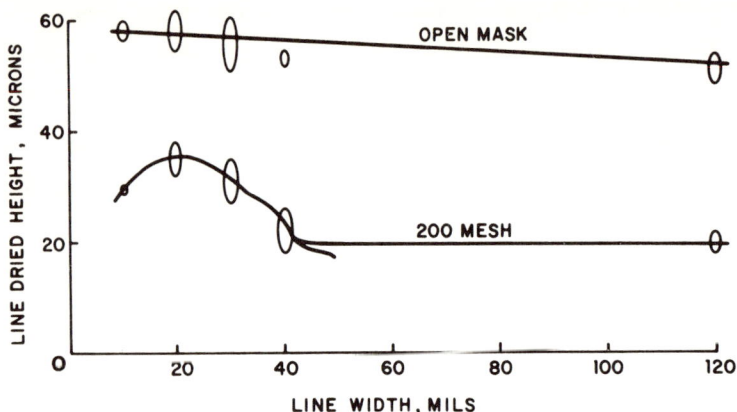

Fig. 15A. "Controlled flow" of paste A through a 200 mesh screen and open metal mask (no mesh) - pattern B of Fig. 2.

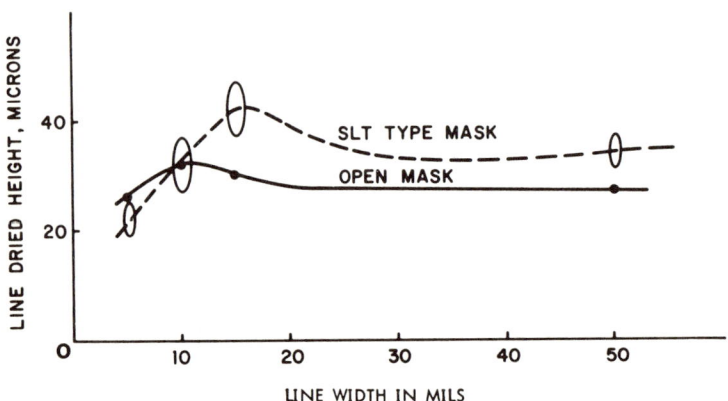

Fig. 15B. "Controlled flow" of paste A through an open metal mask (no mesh) and SLT type mesh mask - pattern A of Fig. 1.

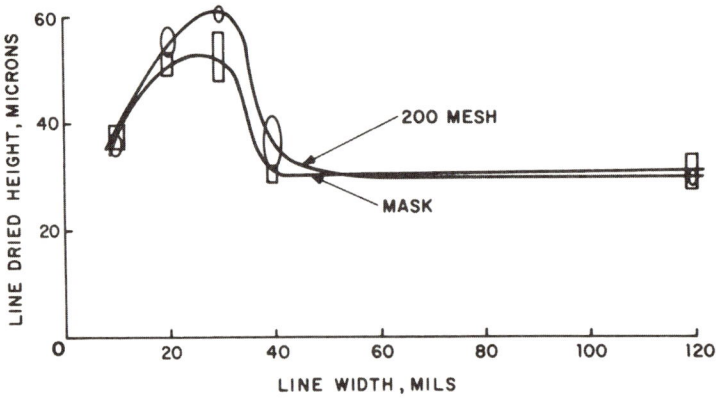

Fig. 16. "Newtonian flow" of paste B through 200 mesh screen and SLT type metal mask - pattern B of Fig. 2.

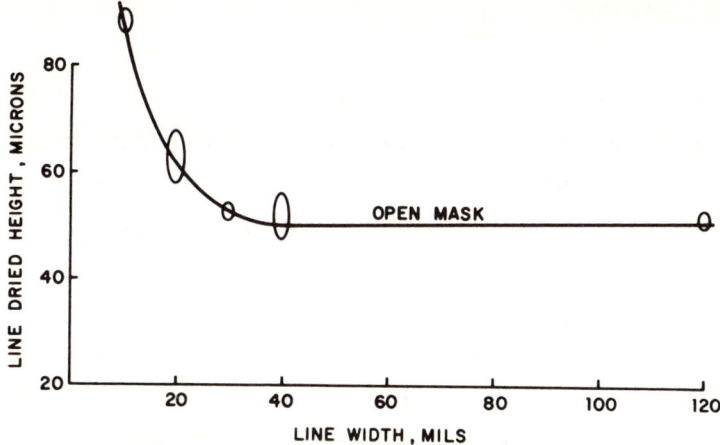

Fig. 17. "Gelled" paste D through an open metal mask (no mesh) - pattern B of Fig. 2.

Fig. 18. Resistor pattern doctored through an open mask.

OTHER PARAMETERS

Although a discussion of other aspects of the screening operation is not the scope of this chapter, a few generalizations can be made. Paste transfer, as measured by the thickness or resistance of the image, in generally increased by: greater squeegee pressure (not greater initial distance of the screen from the substrate), shallower squeegee angle, greater volume of ink in front of the squeegee, and slower speed. It is simple to understand that shallower squeegee angles provide a more vertical force vector on the paste, pushing it more firmly down through the screen. Perhaps it is less simple to see that increased squeegee pressure deforms the squeegee tip to achieve a similar effect. And slower speeds permit more time to totally fill the cavity with paste. Of course, limits apply to all these generalities.

Such experiments should be followed further; these were merely a start, and require confirmation or denial. In particular, they should be extended to production screeners. By doing so, better pastes, masks, and procedures can be designed - - as well as attaining improved predictive and control abilities.

CONCLUSIONS

1. A curve relating image dried thickness to the width of the image apparently exists which contains three distinct sections: an initial linear section (to 15-20 mils width), where the thickness increases with width; a second section (about 20-40 mils width), where the thickness decreases with increasing width; and a third relatively flat section (from about 40 mils width upward), where the thickness remains relatively constant or descends slightly.
2. The slope of the initial linear part of this curve is apparently a function of the screen mesh - - becoming steeper (more change) with decreasing mesh number.
3. The curve also exists for metal masks with mesh. But meshless masks may not show similar characteristics, indicating that the mesh plays an important role in the transfer process.

4. The mechanism of transfer in the three sections of the curve can at least be partially explained in terms of ink immobilization and depression of the mesh to the substrate.

5. Inks with different rheological behavior show radically different curve characteristics, leading to a possible approach to improve thickness scaling characteristics. Pseudoplastic behavior with controlled secondary flow (but not severe gelation) may be better than Newtonian behavior.

6. Knowledge of the curve transition points may aid module design or permit predicting thickness values more accurately or intelligently than before. These phenomena should also be useful for understanding the conductance of lands. It is clear that resistance values expressed in terms of ohms per square or ohms per inch are grossly different for lands of different width - - unlike etched lands.

7. This experimentation is only preliminary and should be followed with additional emphasis on masks (mesh type and no mesh, cavity depth, etc.), paste rheology, and mathematical examination of the volume relationships in the transfer process.

Chapter 2

THICK FILM CONDUCTORS

Before the formulations of individual conductor systems are discussed in detail, it is helpful to take an overview of this field. As shown in Table I, there are many types of conductor systems which are potentially usable. Although some compositions are fluxed (that is, contain relatively small proportions of powdered glass to aid the adherence to the ceramic surface during firing), some materials are capable of bonding directly to the substrate without such a flux. The unfluxed compositions - such as silver:palladium or silver:platinum[1] often require special firing conditions to achieve the mechanical interlock. Chemical bonding is provided by active metal species or the sintered metal process such as those containing titanium, zirconium, or moly/manganese.[2]

TABLE I. Composition Types

* Unfluxed - fluxed
* Powder metal - resinate solutions
* Single metals - alloys
* Fired - resin binder
* Photosensitive binder
* Intermetallics and cermets
* Infiltrated

The metallic constituents of the pastes are commonly composed of powders, but organo-metallics (resinates) can also be used.[3] In general, the resinates, tend to deposit thinner layers than the powders, but mixtures of the two types can often achieve a particularly good result (as shown in Chapter 3). As we can see in Table II, there are a large variety of metallizing systems - such as silver, platinum, gold, and copper, mixtures of powders, and even alloys. As will be

TABLE II. Survey of Thick Film Conductors.

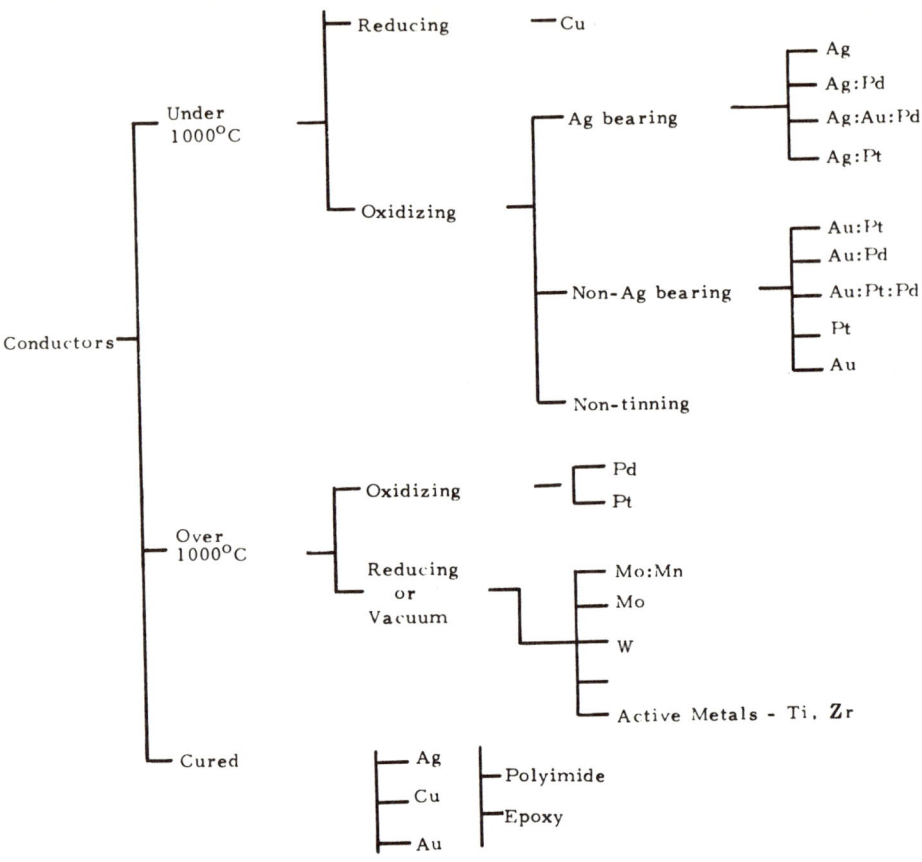

shown in Chapter 4, there are reasons for using alloys of two metals or even three metals to achieve particular benefits. Advances in powder metallurgy have led to the pre-alloying of the metals before the powder is made into a paste, so that greater uniformity can result.[4]

Another interesting subdivision is the differentiation between conductor systems which are fired at high temperatures (generally above 750°C. which drives out all the organic binder and fuses the composition to the substrate) - and those compositions which are cured instead of being fired. The latter typically are epoxy compositions, although polyimides have been described. Generally the fired pastes have greater adhesion, conductance, and stability and are generally preferred throughout the industry, although some epoxy silver compositions are used.

Another technological option which has not seen much commercial usage is the use of a photo-sensitive binder as the vehicle of the paste, so that the paste can be applied completely over the surface of the substrate and then exposed through a mask to delineate a pattern which can then be preferentially developed to form the pattern in the conductor.[5] Along this line, it should be recognized that thick film materials can certainly be put down and fired and then subsequently etched away by photolithographic techniques to produce fine-lined patterns.[6] This type of photographic resolution becomes more important when a screening limits the fabrication of high performance electronic systems, and this approach is generating greater interest for such systems.

Another option is to improve the conductivity of a fired composition by infiltration or capillary soaking with another liquid metal. To do this, the parent pattern should be wettable by the liquid metal to be infiltrated, but not particularly soluble. Thus, such metals as molybdenum or tungsten can be infiltrated with molten copper or silver, and provide highly adherent, highly conductive stripe conductors, particularly within the channels of multi-level ceramics.[7]

Table II represents an arbitrary sub-division of the most notable conductor types into the major classifications of those which fire above 1000°C, under 1000°C, and those which are merely cured (at temperatures from about 100° to 300°C). The firing types can of course be further sub-divided into those which require reducing or inert atmospheres to prevent oxidation of the metallizing, and those which are sufficiently noble to be capable of being fired in an oxidizing atmosphere such as air. It is less costly to fire in air, and furnaces capable of firing at 1000° or less are generally less expensive and less complicated than higher firing furnaces. Fortunately the low temperature air firing formulations also have the greatest compatibility with other thick film passive components such as resistors, capacitors, and dielectric layers. The greatest industrial usage of the higher firing systems has been with moly/manganese. This material provides extremely high adhesion, but must generally be plated with nickel and sometimes gold to permit the joining of other components. Air firing high temperature multi-level ceramics have been described with platinum or palladium metallizing, but this has been less widely used than moly/manganese.

The leading contenders among the lower temperature air firing systems have been silver, gold, silver:palladium, and gold:platinum. The complexity of Table II indicates that there is no material which has universal properties; this implies that there are weaknesses to each system, and indeed this is true.

The figures that follow summarize the major properties of the various conductor metallurgies. As will be shown in subsequent chapters, certain properties are dependent on factors other than the metals - such as the vehicle, frit, and processing conditions, so that this data can be modified. For example, adhesion and conductance can be adjusted rather readily by formulation changes - although with greater facility for some metallurgical combinations than for others.

Table III provides general comments on the major problem with each type of composition. These comments are based on the author's experimental experience and may not represent other formulation types. A few additional comments might be helpful: compositions containing high amounts of gold tend to erode rather readily in normal lead:tin solders, which are commonly used in thick film fabrication. It is difficult to avoid this problem. Any composition containing platinum or any of the platinum metals other than palladium will tend to be extremely expensive due to the raw material cost. However, since even resistor compositions containing expensive metals are finding wider use, this may not be a limitation to some manufacturers since material cost is just a small part of the value of a completed module or circuit.[8] Compositions with large amounts of gold also tend to form brittle intermetallic compounds rather readily with solders - which can potentially give difficulty in terms of stress cracking under adverse conditions.[9]

Table IV describes the adhesion of the various compositions to alumina substrates when they are fired at the recommended conditions. For example, copper must be fired in an inert or reducing atmosphere in order to be tinnable, while the refractory metals such as moly/manganese or tungsten must be fired in the range of 1,500°C. in hydrogen to be adherent and not oxidized. The tensile strength can be arbitrarily sub-divided into three ranges: low (in the order of lower than 3,000 pounds per square inch, which includes copper, silver:

TABLE III. Chief Difficulty for Air-Fired, Tinned Electrodes.

Material	General Comments
Cu	Requires inert or reducing firing properties markedly dependent on particles
Ag	Migration, erosion
Au	Erosion, contact resistance
Ag:Pd	Possible high stress migration
Ag:Au:Pd	Adhesion
Ag:Pt	Cost
Au:Pt	Cost, intermetallics
Au:Pd	Adhesion, porosity
Au:Pt:Pd	Cost, resistance
Pt	Cost, reproducibility
Pd	Adhesion, oxide formation
Mo:Mn	
Mo	High temperature reducing atmosphere
W	Must be plated to be tinned
Ti, Zr	
Cured	Tinning, aging, adhesion, contact resistance

palladium, gold:palladium, and some formulations of silver:platinum, gold:platinum, and platinum. The intermediate range (about 3,000 to 5,000 pounds per square inch) compriese silver:palladium, gold:platinum, and--at the higher end--silver:platinum and gold:platinum·palladium. A specific platinum composition containing a special glass (O. Hommell 03200) produced adhesion in the range of about 5,000 pounds per square inch when fired at 1250°C. The high end of the adhesion spectrum (10,000 pounds per square inch or better) has only been achieved with the active or the refractory metals, or with fritted noble metals cofired with the ceramic at very high temperatures. No data is available for cured systems.

Table V should be self explanatory, and describes the tinnability of each type of land (unmodified by plating). Table VI describes electrical resistance values which have been determined experimentally on laboratory specimens, and correlated with literature--cited values where available.[10] As discussed in

TABLE IV. Adhesion to Alumina (90 - 96%)

Material	Classification	Typical Tensile Strength P.S.I.
Cu	Fair	2400
Ag	Fair	
Au	Poor	
Ag:Pd	Good	3000
Ag:Au:Pd	Fair	2900
Ag:Pt	Good	2000-4500
Au:Pt	Good	3000
Au:Pd	Fair	2000
Au:Pt:Pd	Good	4500
Pt	Fair - Good	2000-5000
Pd	Poor - Fair	-------
Mo:Mn	Excellent	10,000
Mo	Fair	
W	Excellent	10,000
Ti, Zr	Excellent	10,000
Cured	Wide Spread	

Chapter 1, since the amount of paste deposited through a screen or mask can vary according to the line width, this experimental data is cited for a 10 mil wide line to avoid confusion. The sub-divisions into poor to good can be readily observed in this chart.

With this brief introduction into the spectrum of commercially available conductive paste, we can now delve more deeply into individual paste properties and raw material aspects. References 11-13 provide further survey details on thick film conductors.

TABLE V. Unplated Tinnability (10 Sn:90 Pb @ 625°F)

Material	Classification	Main Problem
Cu	Fair - Poor	Oxidation
Ag	Fair - Good	Erosion
Au	Fair - Good	Erosion
Ag:Pd	Good	--
Ag:Au:Pd	Good	--
Ag:Pt	Good	--
Au:Pt	Good	--
Au:Pd	Good	Some erosion
Au:Pt:Pd	Good	--
Pt	Fair - Good	Erratic
Pd	Fair	Oxidation
Mo:Mn	Poor	Requires plating
Mo	Poor	" "
W	Poor	" "
Ti, Zr	Poor	" "
Cured	Poor	

TABLE VI. Screened Electrical Resistance

Material	Classification	Metal	Experimental Ω/.01 inch**	Literature Values $\Omega/\square$
Cu	Good	1.7	.3	
Ag	Good	1.6	.2	0.005
Au	Good	2.3	.6	0.003-.01
Ag:Pd (80:20)	Fair	10.0	1.5	0.015-.04
Ag:Au:Pd (20:55:25)	Poor	--	4.6	
Ag:Pt (80:20)	Fair	17.7	2.0	0.02
Au:Pt (80:20)	Poor	12.3	6.5	0.05-.10
Au:Pd (80:20)	Poor	11.5	10.0	0.015-.12
Au:Pt:Pd (60:20:20)	Poor	--	10.0	
Pt	Fair	10.6	2.0	
Pd*	Fair	10.8	.9	
Mo:Mn (80:20)*	Fair	--	.7	
Mo*	Fair	5.7	.5	
W*	Fair	5.5	.5	
Ti, Zr	Poor	55, 45	--	
Cured	Wide Spread		1.0 up	

* 1500°C.

** 200 Mesh

Chapter 3

SILVER PALLADIUM ELECTRODES

Thick film electrodes generally require the incorporation of three main functional components in the paste: a) a metal or alloy to provide conductance and joinability; b) a flux or frit to adhere the land to the substrate; c) a vehicle to provide screenability. The fired electrodes can be tinned to provide a solder layer for subsequent joining of active devices (chips) by solder reflow, to insure contact of connection pins and, in the case of high resistance electrodes, to provide additional conductance.

This chapter describes a specific paste formulation - one containing 80 parts of silver to 20 parts of palladium - which provides an unusually good compromise of electrode properties.[14] It will show how the formulation and processing parameters produce the final desired properties of the land system. The insights derived from this discussion can, of course, be applied to other metallurgical systems in pastes, and, in fact, some comparison with other systems is included. Before the characteristics of the silver palladium electrode are described in detail, it is beneficial to define the formula, raw material criteria, and production procedure.

TABLE I. Typical Formulation (Parts by Weight).

56.4	Silver Powder } Metals		
14.1	Palladium Black		75% Solids
3.0	Bi_2O_3 } Flux		
1.5	Borosilicate glass		
3.6	Ethyl cellulose - resin		
1.6	2-furoic acid - flow control agent		25% Vehicle
17.3	Butyl carbitol acetate-solvent		
2.5	Igepal C0430* - surfactant		

Materials

Silver. The choice of silver powder is dictated primarily by considerations of purity and shape. Since it is the main ingredient, the surface area should be relatively low to permit the high solids loadings which lead to high conductivity. Flake silver tends to produce gelation, so that powders are preferred. Particles from about 1 to 5 microns average, with surface areas of about 0.5 to 1.5 square meters per gram and an apparent density (Scott volumeter) of about 5 to 16 grams per cubic inch are quite acceptable. Such silver powders are readily available commercially, but an acceptable silver powder of this type can be prepared by reducing dilute silver nitrate solutions with 1:1 aqueous hydrazine solution at room temperature. Silver powders with excessively large particles can tend to clog screens, erode in solder, produce poor adhesion, and tin poorly. In such cases, poor tinning can be attributed to the presence of palladium-rich areas which oxidize during firing, since there are insufficient small silver particles to completely surround them. (Oxide surfaces do not tin.) As will be seen later, this avoiding of oxidation is one of the primary considerations to be controlled in the performance of an electrode of this type. However, the presence of oxides in the starting metals does not inhibit performance characteristics if they do not interfere with the formula ratios or grossly reduce the metal content.

Palladium. The palladium powder particle size and surface area strongly affect the performance of the electrode, perhaps much more so than the silver, even though the palladium is present in much smaller quantities. Large particles can result in slightly higher resistance and poor adhesion. Thus, an acceptable surface area from about 10 to 25 square meters per gram is suggested for the palladium powder. Evaluations have also shown that raw or sintered palladium powder as well as palladium oxide can be used in this formulation with little sacrifice of properties if the surface area criteria are met. Of course the particle surface area and size not only affects the fired electrode performance, but also is a controlling factor on rheology and screening. Larger particles produce greater flow, and if they are not too large to pass through the screen, sometimes less screen clogging. This seemingly confusing statement cen be explained in terms of vehicle immobilization on the pigment surface areas. The pigments must either be wetted by vehicle or resolve their surface activities by agglomerating with nearby particles. With pigment loading close to the oil absorption point - the amount of vehicle needed to wet the powder into a just barely cohesive mass - which describes the inks under discussion - these phenomena are rather drastic. Thus smaller particles use considerably more vehicle to wet the pigment surface, leaving less vehicle for flow and viscous phenomena. This can explain why screen clogging can occur from inks or pastes made from very small particle palladium powders. It is not particularly difficult to eliminate this potential difficulty; rheological optimization can be achieved by adjustment of the vehicle resin content, the solids loading in the ink, and usage of surface active materials. The above formulation can produce non-clogging silver palladium electrode pastes which screen rather well.

<u>Flux</u>. Although mixtures of Bi_2O_3 with glass generally seem to permit good tinning,[15, 16] pure Bi_2O_3 does not provide adequate adhesion. The only glasses which seem to provide both good tinning and adhesion are relatively low melting lead borosilicate glasses. The following considerations probably apply here: the frit is dispersed like raisins in a cake throughout the paste mixture, bonding both to the ceramic substrate and to the metal powders, which helps to create a tough composite. It is believed that the land strength is a more predominant function of the cohesive properties of the electrode; the frit-to-ceramic bond is quite strong and probably not the weak link. When the lands are tinned, any non-removed frit on the surface prevents land wetting by the liquid solder. The soldering flux is a mild reducing agent which is capable of reacting with and removing basic metal oxides. The lower melting, less stable, or more basic oxides which might be included in the glass, (as PbO, Bi_2O_3, V_2O_5, CdO, Na_2O, etc.) can be partially reacted and removed in this process while stable refractory oxides with high free energies of formation (as Al_2O_3, SiO_2, B_2O_3) cannot be removed in this manner. Thus glasses with more of the former and less of the latter should tin better, as experimental evidence supports. Unfortunately, the latter provides structural strength to the glass (they are the common "glass formers"), and are necessary to hold the composite together so that compromise may be necessary.

Tensile failure may be viewed to be a predominantly cohesive failure of the lands. During firing, the frit in the lands penetrates into the glassy phase between the alumina grains of the substrate, and possibly some back diffusion of that flux-glassy phase penetrates up into the land. This leaves a slightly glass-deficient region somewhat above the original ceramic surface; this depleted region is now the weak link in the system. However, for some electrode system - such as gold-platinum, it is not much weaker than the remainder of the lands - perhaps only a few hundred p.s.i. less. After the land is pulled from the substrate, such as in a tensile test, a glassy phase with some trapped metal particles is left on the substrate. These observations - both the flux penetration into the substrate and the remaining glassy film - have been made by electron microprobe.

Anything which interferes with the frit-to-metal or metal-to-metal particle bond system should severely degrade adhesion. Thus it is preferable that the frit expansion coefficient should more closely match the metals than the ceramic substrate. The frit softening point should be relatively low, (between about $450^{\circ}C$ and $600^{\circ}C$), to provide good metal wetting at the chosen $750^{\circ}C$ firing. Also the glass viscosity at firing should be low enough to provide good capillary soaking of the metal particles, and, coupled with a high specific gravity, flow downward toward the ceramic to support the base of the electrode. The top of the electrode does not require such support, since the solder serves that function.

Vehicle. The liquid vehicle which carries these powders for printing purposes should be as stable as possible. One aspect is volatility; low volatility solvents as used in this formula do not dry out rapidly during screening or condense on the inside of the jar. However low volatility solvents add the additional problem of preventing excessive flow since the solvent is not lost rapidly after screening in normal ambient conditions. This problem be overcome with several types of flow controlling additives, but some of the common gelating agents used in ink technology are not applicable here: often the drying temperature exceed the gel breaking temperature, and some gelation agents (as colloidal silicas) do not pyrolize during firing, and thus interfere with the tinnability of the resulting land. Thus a new class of flow controlling additives was developed to optimize this type of paste. Materials which work particularly well are those that sublime at an appreciable rate below their melting point. For not completely explained reasons, these materials - which include furoic acid, terephthalic acid, etc. - can completely eliminate secondary flow of even relatively thick deposits and thus maintain fine line definition without noticeably affecting the rheology at higher rates of shear. Another group of materials which work quite well are inorganic ionic solids which apparently create a very slight gelation at low rates of shear. Particularly useful materials of this type are ammonium sulphate and sodium chloride. However, the latter might be suspected to leave a potentially corrosive halide residue. Furoic acid (mentioned in the above formulation) also acts as a wetting agent for the metal powders and permits very sharp printing.

Ethyl cellulose is used to provide the viscosity of the inks. It provides high viscosity and limited gelation with rather low concentrations in medium polarity solvents. Again a compromise is called for in choosing the molecular weight and degree of ethoxylation. Excessive molecular weights cause gelation and total ethoxylation limits solubility in polar solvents such as butyl carbitol acetate. A good choice is a low molecular weight such as Hercules grade N11 or N22. Experiments have shown that the fissuring of the edges of very fine lines can be reduced with the use of lower molecular weight resins. Thermogravimetric analysis indicates decompostion can begin as low as 200°C in air.

Since these electrodes are designed to be used with glaze resistors, it is desirable that the formulation contain no superfluous metallic or corrosive ions. This rather restricts the choice of wetting agents. Furthermore, the vehicle should be relatively incompatible with water to avoid changes in rheological properties with humidity variations, as well as being nonreactive. For these reasons, a particularly applicable surfactant is the non-ionic hydrophobic surfactant of the chemical type nonylphenoxy polyoxyethylene ethanol, which provides both good wetting and lubrication. The surfactant concentration can be varied to adapt the basic ink to special screening requirements, acting as an internal lubricant, without affecting the performance of the electrode in any discernible manner. During firing, the entire vehicle is pyrolyzed and is driven from the electrode.

Metal Ratio. The ratio of 80 parts silver by weight to 20 parts of palladium has been experimentally verified to be the best compromise for certain types of thick film operations. Higher silver contents accelerate both erosion in solder and potential migration tendencies, while lower silver contents do not tend to tin as well, particularly with relatively slow firing cycles. For example, x-ray data showed some palladium oxide formation in 65 silver:35 palladium ratios after firing at 750° for about an hour, but not in 80 silver:20 palladium formulations, even with very slow cooling cycles. This is related to the very high partial pressures of oxygen required for oxide formation in higher silver:palladium alloys, to satisfy the free energy requirements for the oxidation. In the firing cycles typical to those recommended for silver:palladium oxide glaze resistors, there is extensive experimental evidence to verify the above. (Properties of AgPd can be found in references 17 - 27.)

TABLE 2. Ag:Pd Ratios.

Ag:Pd Ratios	Wetting by Solder[1]	Appearance	Erosion[2]	Resistance[3]	Adhesion (Estimated)
90:10	Good	Good	Excessive	0.45	F-G
85:15	Good	Good	Some	0.54	F-G
82:18	Good	Good	Minor	0.60	F-G
80:20	Good	Good	None	-	F-G
77:23	Good	Some dull areas & pin holes	None	0.81	F-G
75:25	Good	Some dull areas & pin holes	None	0.81	F-G
70:30	Poor	Mottled surface, small fissures	None	0.90	F
65:35	Poor	Mottled surface, small fissures	None	1.20	F

1) 5 second dip
2) 3-5 second dips
3) ohms/inch/15 mils (400 mesh)

Electrodes with high palladium ratios tend to show uneven solder surfaces, indicating some oxide formation during firing. Very low palladium contents tend to show some etching (erosion) in lead tin solders. Unexpectedly, silver bearing solders do not seem to reduce this erosion tendency. Furthermore, a main culprit is the tin in solders: higher tin contents lead to considerably greater erosion. The conductivity of the electrodes tends to increase at the high silver end as does the tendency toward silver migration. Of course, the cost decreases with decreasing palladium content.

Paste Preparation. To insure homogeneity, the dry metal and flux powders can be shaken together before being combined with pre-mixed liquid components (the vehicle). Of course during this dry mixing process, care should be taken to avoid segregation of the powders. Furthermore, during the vehicle preparation, excessive heat should be avoided to minimize polymerization or chain scission which might lead to rheological changes. It is also possible to use pre-alloyed silver palladium powders in the proper ratio to enhance homogeneity of the system. Such alloy powders can be directly deposited from solution. Furthermore, if desired, the flux ingredients can also be pre-reacted; as far as laboratory tests have been able to determine, pre-reacted bismuth oxide glasses work just as well as when the indivdual components are mixed together.

After the powders have been mixed with the vehicle and are wet out, sufficient passes are applied on a 3 roll mill to provide adequate dispersion and reduction of pigment agglomerates. This milling is not normally considered to be an attrition process; it mainly pulls agglomerate units apart by shearing forces which are transmitted through the vehicle as the rolls turn.[28] Actually clearance at the roll nip is usually considerably larger than the resulting particle size. To achieve uniformity of paste production, adequate pressure controls and roll cooling facilities should be applied.

In order to avoid possible screening problems from clogging by excessively large particles, all powders should be capable of passing through at least 325 mesh before being used in the paste. Then the milling of the ink should be controlled to produce grind gage readings far smaller than the mesh size.

Electrode Characteristics. Now that we have looked at some aspects of this formula, the properties and characteristics can be discussed in more detail. A good electrode system for thick film circuitry should have the following properties: good screenability and flow control; form dense nonporous lands; show good adhesion to the substrate both in tension, shear, and compression; have rheological stability; show high conductivity, both tinned and untinned; show no adverse metallurgical interaction with either active or passive devices; be completely compatible with glazed resistors with respect to such properties as TCR, drift, interfacial hot spots, scaling, etc.; provide excellent tinnability; and above all, be reliable and reproducible. All of these properties, if possible, should be obtained at relatively low cost. As the following discussion will show, this electrode system satisfies these criteria to a very large extent. Of course, every definition of acceptability is related to a context and adequacy in one circuit might be absolutely unusable in another. For example, adhesion which is good enough to reliably support an active device weighing several milligrams might fail as a lead attachment subject to several pounds pull.

Unless otherwise noted, the following processing conditions were used in defining the electrode characteristics:

Screen	325 mesh, approximately 1 mil emulsion thickness
Substrate	Alsimag 614, 96% alumina (American Lava Co.)
Drying	90°C - 100°C - 15 minutes
Firing	750°C, either belt furnace (BTU) at 1-1/2 inches per minute, or CM furnace for 25 minutes in, 15 minutes soak, 25 minutes out (both acceptable resistor firing cycles).
Tinning	5 - 10 seconds in 10 Sn:90 Pb solder at 620 - 630°F; fluxed with Alpha 100 - 40 rosin flux.

All other processing steps, as reflow, resistor application and firing, etc. are standard thick film procedures.

Morphology. The firing process not only adheres the metal and flux to the ceramic substrate, but also sinters them into a coherent deposit. The silver: palladium lands sinter and fuse exceptionally well from 600°C to over 1000°C, forming a very dense, uniform land compared to other metallurgical systems. A comparison of a commercial gold-platinum paste to a fired silver-palladium paste can be readily be made from the micrographs of the crosssections in Figs. 1 and 2 and from the micrographs of the top surfaces of lands illuminated with transmitted light in Figs. 3 and 4. The improved denseness of this formula is desirable to limit the surfaces available to contact by solder, since a solderable material is always dissolved to some extent in the molten solder. The less the contact area, the less the erosion and resulting effect on resistance. In addition, conductivity most closely approaches the bulk value if all particles are intimately touching and the density of the conductor is at a maximum.

Fig. 1.   Cross-section of tinned Ag:Pd land.

Fig. 2.   Cross-section of Au:Pt land.

Fig. 3. Top view of Ag:Pd transmitted light.

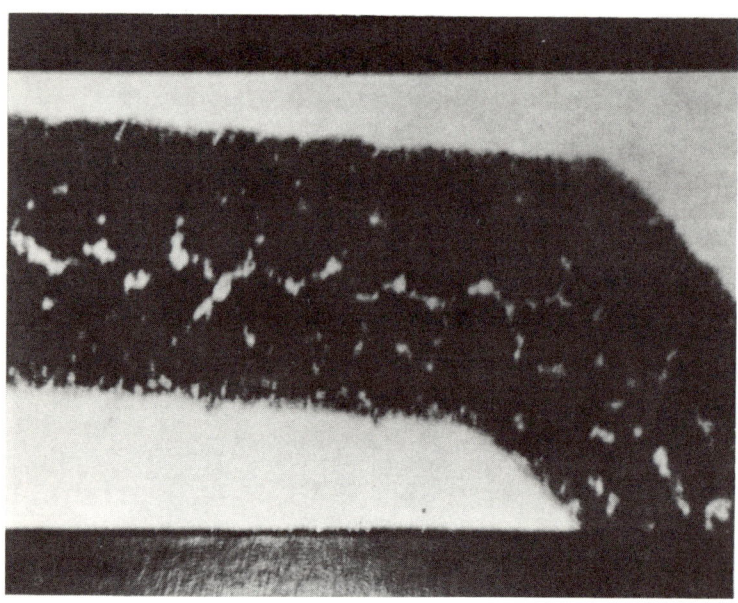

Fig. 4. Top view of Au:Pt transmitted light.

As the individual particles of the paste sinter and coalesce during firing, the volume of the material is reduced and this sometimes leads to shrinkage and even fissuring of certain types of lands. The distribution of particles in the paste is particularly important here. Excessively rapid drying of volatile vehicles sometimes fails to permit a uniform settling of pigments in an even layer, creating incipient or even gross cracks during firing. However this formula contains not only a slow drying solvent but also a wetting agent and a flow control additive (such as furoic acid) which inhibits the formation of these cracks. In fact, even drastically rapid drying (as, for example, at 300°C), does not create cracks after firing. It is interesting to note that these concepts are also successful when applied to gold-platinum formulations; 5% terephthalic acid or ammonium sulfate has been shown to consistently reduce cracking and porosity of gold-platinum paste. As Figs. 5 and 6 show, a gold-platinum formula made with a similar vehicle did not crack when compared with a gold-platinum formula without a flow control additive when both were sintered and fired on the same screening production equipment.

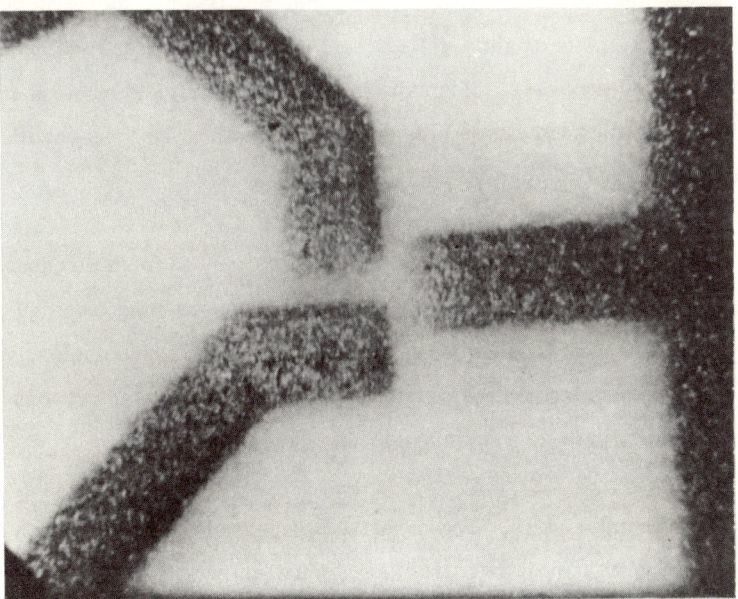

Fig. 5. Top view Au:Pt with improved vehicle.

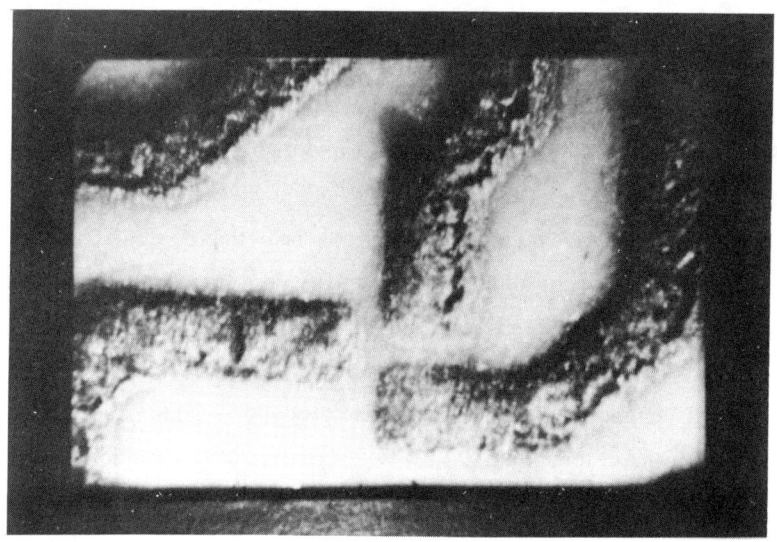

Fig. 6. Top view Au:Pt - no additive.

The profile of screened films tends to be quite rounded if the ink has considerable flow, low solids, large particles, limited particle wetting, and good wetting of the printed surface. On the other hand, profiles are more rectangular or trapezoidal if the ink is slightly gelled, has a high viscosity and yield value, and rapid thixotropic recovery. Rounding is not particularly desirable if it creates thin edges which run beyond the desired dimension, tin less readily than thicker areas, or erode in solder.

The rheological characteristics of these pastes tend to form a reasonably rectangular profile with a minimum thin edge if proper screening techniques are used. However, excessive squeegee pressure, low squeegee angles, very slow screening speeds, inadequate contact of mask and substrate, or worn emulsion or mask edges can of course, create thin edges with any ink.

Strength Characteristics. The cohesive strength of these electrodes is quite good as far as air fired, relatively low temperature metallizing is concerned - probably exceeding 5000-6000 psi in tension. The compressive strength - represented by the amount of pressure on the land required to cause a land rupture - is extremely high: some tests have shown resistance to 150,000 psi without degradation. In addition, active devices joined to these tinned lands[29] (as in the SLT process) show strengths which commonly are equivalent to 100 grams per pad. Tensile adhesion can be directly measured by pulling a flat headed rivet which has been solder reflowed to a dot of the same size of fired electrode paste (0.150 inch diameter) directly upward at a speed of 0.02 inches per minute on an Instron tester.

Of course, the adhesion of any electrode system is dependent both on the processing conditions and the type and amount of flux used in the formula. There is a distinct tradeoff in the amount of flux necessary for good adhesion and the interference of that flux with land tinnability. Not all these relationships tend to be predictable; Fig. 7 and Table 3 show that a particular borosilicate frit actually shows a maximum of adhesion at a particular concentration (which is entirely unexpected). Furthermore, certain fluxes tend to interfere with solderability less than others. The two part flux (Bi_2O_3 and Drakenfeld 1527 lead borosilicate glass) provides good adhesion in this metallurgical system without sacrifice of tinnability. These materials seem to behave synergistically - the combination is far better than each alone. The adhesion of even the unfluxed silver-palladium electrode seems to be quite good when fired above $750°C$ and particularly so at $1000°C$. In this case, adhesion increases with higher firing temperature, although this is not universally so for all electrode systems. Unfortunately, the adhesion of the unfluxed lands decreases considerably after tinning, so that high adhesion requirements might call for glass frits or fluxes. Some applications preclude the use of Bi_2O_3 or other semiconductor oxides due to doping effects on resistors or capacitors. For such uses, glasses similar to Drakenfeld E1313 lead borosilicate have been found to be particularly effective in relatively low concentration to provide a reasonable compromise of tinnability and adhesion. The maximum adhesion value with the E1313 glass, which occurs at about 4% frit by weight, may be related to the strength of the solder bond on the electrode surface which becomes weaker with very high flux concentrations. This maximum value is quite fortuitous, since beyond about 4%, the tinnability degrades somewhat. Thus an electrode paste with 3 to 4% of this glass as a binder is an acceptable compromise for some conditions.

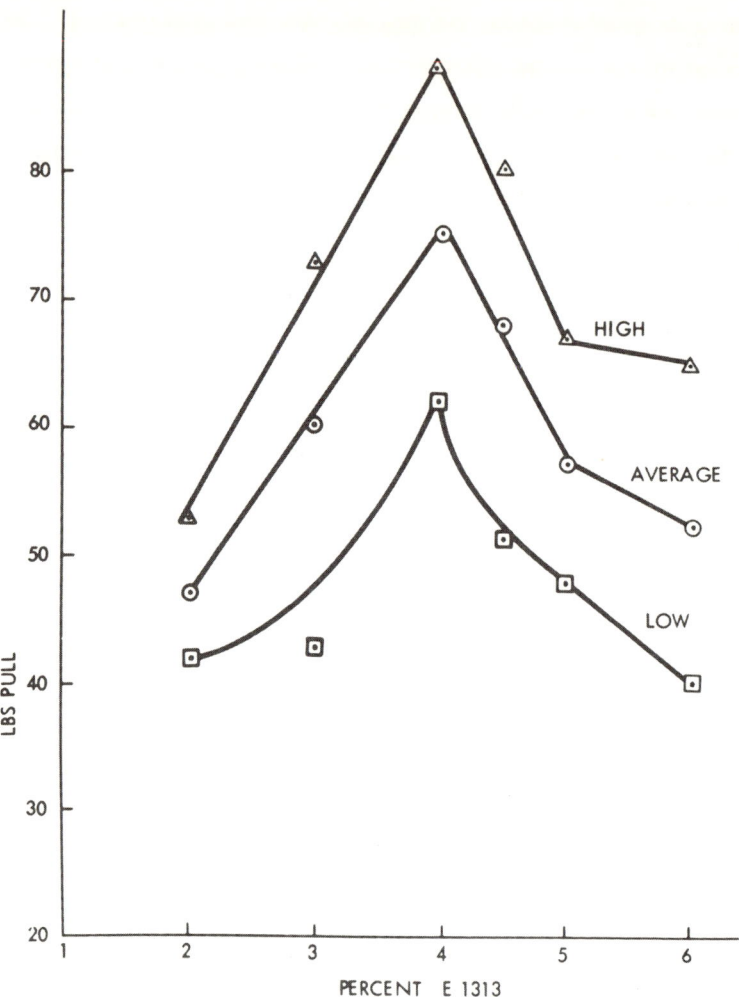

Fig. 7. Effect of E1313 glass concentration on adhesion of 80 Ag:20 Pd electrode–0.150" rivet.

TABLE 3. Comparison of the Adhesion of Different Flux Concentrations.

Flux	Adhesion (lbs. pull)		
	Low	High	Average
4 Bi_2O_3, 2-1527	15	60	37-40
6 Bi_2O_3, 3-1527	20	60	41
8 Bi_2O_3, 4-1527	30	65	50
2 E1313	33	48	40
3 E1313	43	73	60
4 E1313	62	88	75
4.5 E1313	51	80	68
5 E1313	48	67	57
6 E1313	40	65	52

As with other paste systems, the adhesion of silver palladium electrodes tends to degrade after either thermal cycling or high temperature storage.[30] Surprisingly, silver:lead:tin eutectic solders do not improve either the adhesion characteristics or the resistance to loss of adhesion upon such accelerated aging tests. Furthermore, the higher the tin concentration the more metallurgical interactions occur and the lower the strength tends to be, both before and after stressing, Table 4. As with other fired paste systems, the adhesion of the silver palladium electrode can be improved on glazed surfaces. The unfluxed silver palladium bonds well to certain glazed surfaces, and is quite tinnable in such a format.

TABLE 4. Adhesion Pull Values (lbs.) Before and After Stressing, with Two Different Solders.

Paste		Solder							
		10 Sn/90 Pb				3 Ag/27Sn/70Pb			
		Change After Stress	Low	High	Avg.	Change After Stress	Low	High	Avg.
Au:Pt	4 controls, time zero		44	59	52		17	31	25
	5 controls, 944 hrs room temp.	70%	43	47	44	61%	30	36	31
	150°C, 944 hrs aging (10)		25	38	31		11	24	19
Ag:Pd Bi$_2$O$_3$, glass	4 controls, time zero		50	70	58		30	38	33
	5 controls, 944 hrs room temp.	67%	55	60	58	0%	28	40	33
	150°C, 944 hrs aging (10)		29	47	39		27	39	33
Ag:Pd 4% E1313 glass	4 controls, time zero		53	65	58		36	43	40
	5 controls, 944 hrs room temp.	77%	46	70	58	58%	34	45	39
	150°C, 944 hrs aging		41	49	45		14	31	23

<u>Rheological Stability and Reproducibility</u>. Non-volatile vehical formulations of this type are quite stable if certain chemical interactions do not occur. Pastes which do not contain basic metal oxides (such as Bi_2O_3), have shown jar stability for several years. However if such materials as furoic or terephthalic acids are used as secondary flow controlling agents, interaction with basic metal oxides can occur, and in such cases, the viscosity can increase to an unusable level after four to six months storage. No such interaction has been observed with ammonium sulphate as a flow stopper. Typical viscosity trends of paste of this type are shown in Fig. 8. The viscosity increases of reactant systems can be minimized by refrigerating the pastes, replacing the reactant flow stoppers with non-reactant materials, or by changing the flux. The increase in viscosity with aging becomes less severe with better paste dispersion. Pilot production lots made on large mills tend to be considerably more stable than laboratory

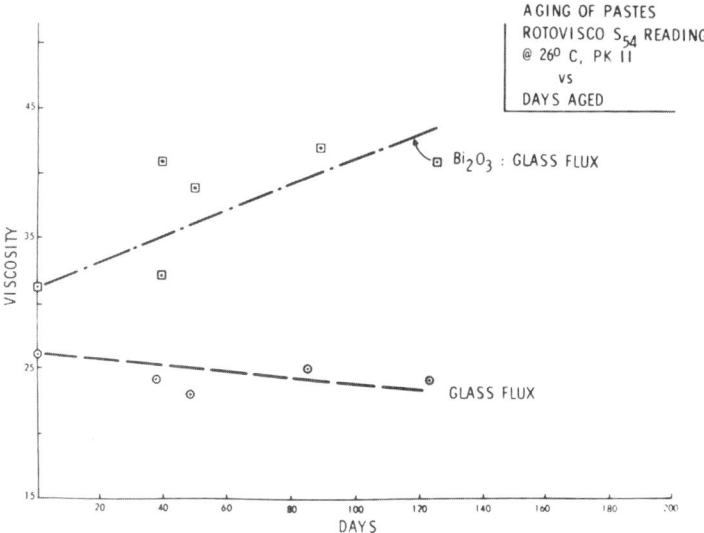

Fig. 8. Aging of pastes.

inks milled with very small rolls. Careful material specifications and close control of production procedures have attained high reproducibility of paste properties.

Resistance. Resistance of any electrode paste system can be expressed in many ways, but it has been found particularly convenient to express it in terms of ohms per inch of lines of specific width. Since it has already been observed that the profiles of screened lands are not generally rectangular, and since lands of different widths - even though screened at the same time through the same screen - will generally have different heights, it is often fallacious and misleading to refer to the resistance of these lands in terms of ohms per square. The resistance value which results from a particular screening operation is dependent on a series of factors which may not be reproducible unless the same operating conditions are maintained. Thus geometric factors are pertinent as well as the resistivity of the materials; the total electrode structure must be taken into consideration - the screen, screening conditions, firing and other treatments, etc.

However, if materials are compared under the same conditions (as with adhesion), valuable information may be gained. Sufficient samples should always be taken to limit the importance of chance perturbations of the printed image - as clogs, irregular squeegee pressures, uneven substrates, etc. As a gross comparison, the conductivity of the silver palladium land described here is about seven to ten times greater than gold-platinum pastes. Table 5 compares the resistance of these two systems.

Maximum conductivity of a paste system can, of course, be achieved with pure, highly conductive metals such as gold, silver and copper. As Table 6 shows, the silver palladium electrodes attain about 5 times the minimum attainable resistances of those pure metal systems.[20,31] The 80 silver:20 palladium alloy has a resistivity of about 10 microhm centimeters while the gold, silver, and copper are about 2 microhm centimeters. The good sintering of the silver palladium alloy enables a close approach to bulk resistivity (somewhere between 1.5 and 2.5 times bulk for lines which are wide enough to minimize the contribution of printing defects).

TABLE 5. Resistance of Ag:Pd and Au:Pt

	Resistance, Ohms per Inch, 200 Mesh		
Material	5 mils wide**	10 mils wide**	15 mils wide**
Ag:Pd	4.8 - 6.7	1.4 - 1.7	0.8 - 1.0
Ag:Pd Tinned	5.5 - 7.2	1.0 - 1.1	0.3
Au:Pt	6.5 - 8.5	8.8 - 9.5	4.9 - 5.3
Au:Pt Tinned	5 - 8.7	1.2 - 1.3	0.4 - 0.5

**Ten samples for each condition.

TABLE 6. Electrical Resistance Comparison.

Electrode Type	Metal Resistivity Microhm-cm	Land Resistance 200 Mesh ohms/inch/10 mils	Metal TCR (20-100°C)
Gold	2.4	0.3	4×10^{-3}
Silver	1.5	0.2	4×10^{-3}
Silver-Palladium	10	1.5	0.5×10^{-3}
Copper	1.7	0.3	4×10^{-3}

There are interesting relationships between the resistance of tinned and untinned lands: for electrode system with reasonably high resistivity - such as gold-platinum, the solder layer conducts a sizeable proportion of the current and therefore is a major factor in the resistance measurement of the tinned land; however, since the value of the resistivity of 10 tin:90 lead solder is in the same range as that of the silver palladium alloy, each layer is a significant factor in the measurement. Typically, therefore, tinned gold-platinum electrodes improve resistance by an order of magnitude after tinning, while the relationship for silver palladium electrodes is not so clear. A surprising switchover between the conductance of tinned and untinned silver palladium lands is experimentally observed. When screened through 200 mesh, 5 mil lands often increase in resistance after tinning, while 10 mil lands improve about 30% or so, and 15 mil lines reach about half the original resistance after tinning. This can be explained by two factors: the amount of land eroded by the solder compared to the amount of land originally present, and the supportive conductance provided by the solder. Intermetallics and alloying seem to be second order phenomena here. Narrow thin lines lose more conductive material than they gain back from the very thin solder which is deposited. Wider thicker lands receive more solder and are eroded proportionally less. However this relationship can change drastically when the thickness of the lands are increased, so that experimentally 6 mil wide lines deposited thickly through metal masks have been shown to actually improve conductivity after tinning. The important thing to be recognized here is that very often generalizations do not hold up on actual experimentation, and that an understanding of the factors and parameters involved in the measurement and format are necessary before prediction can be made.

The TCR (temperature coefficient of resistance) of the untinned lands is reasonably low (roughly 2.5 to 4% change in the temperature range of 25°C - 100°C), but the tinned lands are really the matter of concern, since they are normally used in module fabrication. Here the particularly high conductivity of these lands becomes beneficial - since a large part of the current travels through the land instead of the solder, the TCR of the composite is considerably lower than with less conductive lands. A few non-precise checks of the TCR of tinned silver palladium lands showed 7 to 11% change over the above temperature range, while tinned gold-platinum electrodes showed about 18 to 19% change. This might be important in circuits where low value resistors are connected with long leads; in such cases, the land resistance can be a significant portion of the total resistance, and changes due to temperature variations can be troublesome.

The resistance of the fired lands is not appreciable changed by the amount of milling; a lot of ink was split into four milling groups - 10, 20, 30, 40 passes each. There was no statistically significant difference in either the tinned or untinned resistance values. Furthermore half of two of the groups - 20 and 40 passes - were dried for 2 hours in air instead of normal 15 minutes at 100°C to examine the effect of drying conditions. Again, no significant difference in resistance was observed, (see Table 7). Thus, if land compaction is a factor in solder height and resulting resistance values, dispersion and drying do not appear to be controlling parameters. This may be due to the equallizing effects of the gelation and thixotropy, which may tend to hide pigment settling phenomena in these formulae. Surprisingly, there is not much improvement in conductance when paste solids are increased from 75 to 80%. Much greater improvements can be obtained by depositing thicker lands (such as with wider meshes, thicker emulsions, or deeper etched metal masks).

TABLE 7. Resistance Values (Ω/0.270"/0.006") of Various Milling Groups and Formula Variations.

Ink	Untinned Ω		Tinned Ω	
	Spread	Average	Spread	Average
10 passes	0.81-0.92	0.86	0.62-0.88	0.79
20 passes	0.78-0.88	0.84	0.68-0.95	0.78
20 passes (2 hrs air)	0.85-0.91	0.87	0.60-0.90	0.80
30 passes	0.81-0.92	0.87	0.65-0.88	0.76
40 passes	0.73-0.85	0.80	0.63-0.85	0.73
40 passes (2 hrs air)	0.80-0.85	0.82	0.79-0.90	0.84
Unfluxed Ag:Pd	0.78-0.89	0.83	0.50-0.60	0.55
7553 (thickened) (2 hrs air)	3.9-5.2	4.3	0.53-0.73	0.65

The solder thickness on these lands tends to be somewhat less than on gold-platinum lands. If this is due to the gold-platinum lands holding more solder due to the sluggishness of the gold-loaded solder (which does not run off so rapidly), a substitution of some gold for the silver in a silver palladium formulation should provide a thicker solder layer. A replacement of 10% silver with gold showed initially interesting results: the lands tinned quite well and evenly, and decreased in resistance after tinning considerably more than the similar formulation without the gold - implying greater solder height, see Table 8. The untinned lands showed higher resistance than the non-gold bearing paste, while the tinned lands were considerably less resistive. Obviously further investigations are required for a complete understanding of this situation. It should be noted that the presence of gold can lead to the formation of harmful intermetallics with either solder or with copper bearing materials which might be joined to the solder in subsequent module fabrication procedures. The absence of such gold intermetallic formation is an important advantage of the silver-palladium system.

TABLE 8. Effect of Gold Substitution on Resistance.

| | \multicolumn{6}{c}{Resistance, Ω/in. 200 mesh} |
| | \multicolumn{6}{c}{750°C 2"/minute firing} |

Paste	Untinned			Tinned		
	5 mils	10 mils	15 mils	5 mils	10 mils	15 mils
Typical Ag:Pd 4% E1313 glass	5.1	1.5	0.8	5.5	1.0	0.5
(10% Au)	5.4	1.8	1.0	3.4	0.75	0.44

Resistor Compatibility. Obviously, compatibility with glazed resistors and other passive components is a prime aspect of the performance of thick film electrodes. Our investigations dealt with three main resistor systems: DuPont glazed resistors, IBM low resistance doped resistors,[32] and Indium Oxide resistors.[33] Excellent compatibility has been shown with all three resistor systems. Prevously, it was noted that low resistance glazed resistors (in the order of 100 ohms per square and lower) do not scale well with electrodes containing large amounts of gold. Scaling is the ability to reproduce a consistent ohm per square value with no dependence on the resistor size or resistor-electrode contact area. If a high contact resistance is formed between the electrode and resistor, scaling can be severely perturbed (the resistance value depending largely on the contact resistance). However, scaling is quite good with such resistors on silver palladium contacts. The contact resistance of silver:palladium oxide glazed resistors is generally lower on these electrodes than on gold-platinum electrodes. High current pulses tend to burn out the gold-platinum/resistor interface, but when the resistors are applied on silver palladium electrodes, such pulses burn out the resistors in the center, away from the contact area. A considerable amount of resistor testing for long periods showed excellent compatibility, contact resistance, and drift characteristics.

Drift and TCR with the silver palladium electrodes is about equivalent to that with platinum electrodes for medium and low resistance resistors. Nominal resistance values, data spreads, and interactions also are equivalent. Drift with silver palladium electrodes tends to be somewhat better than with gold-platinum electrodes with DuPont 7827 and 7828 resistors. Scaling is not linear however, and small resistors tend to be somewhat lower in value than expected. However, this is a predictable phenomenon and can be designed into the circuit layout. Furthermore, nominal resistor values are lower than with gold bearing electrodes - again in a predictable manner - from about 15 to 30% lower for DuPont 7827 resistors. When bismuth oxide flux is used in addition to borosilicate glasses in the electrode, resistor values are very slightly higher and scaling is very slightly better.

When resistors are overlapped a considerable distance on the electrode, sometimes cracks appear at the interface.[34] Electrical tests have shown that these cracks have no effect on the resistor values and performance, although severe cases could possible create electrical opens. Thus, although this cracking seems to have no significant practical effect, it can be minimized by minimizing the overlap region of the resistor on the electrode. An overlapped region of 5 mils is preferred. The cracking has also been shown to be related to the silver in the electrode; higher palladium ratios produce less cracking. Figure 9 compares the resistance across a DuPont 7800 series resistor when overlapped 10 mils on both Au:Pt and Ag:Pd electrodes.

Other Passive Components

Capacitors. It is not normally considered desirable to use any silver bearing electrode in the fabrication of screened capacitors, since the combination of thin dielectric layer and relatively high DC voltage are ideal circumstances for silver migration to occur. Severe stress tests on such capacitors (in rain

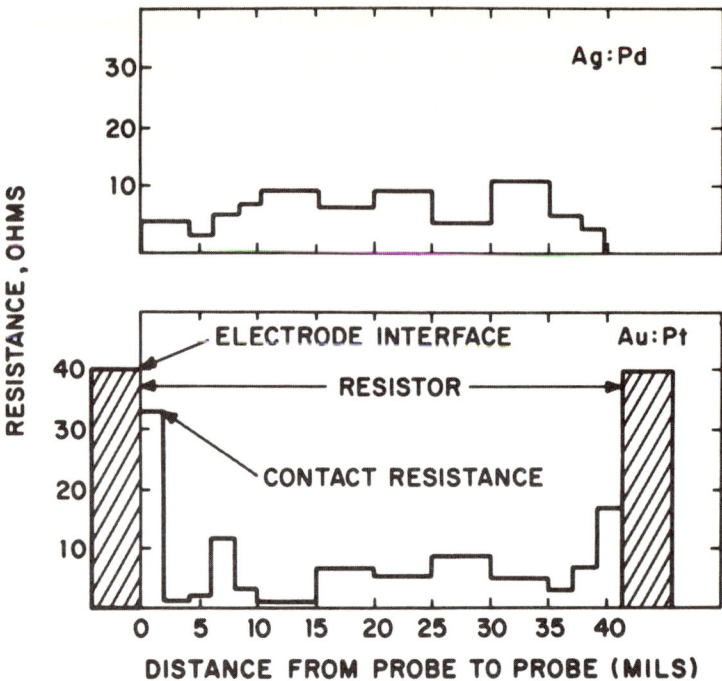

Fig. 9. Incremental resistance measurements across Ag:Pd and Au:Pt interfaces.

chambers at 100% relative humidity) have consistently shown failure due to shorting when silver bearing electrodes were used, even when organic overcoatings were used as encapsulants. It would seem that if silver were used in the negative electrode only, no migration would occur due to the absence of positive silver ions from the positive electrode. A quick test indicated that this format did indeed hold up better than the opposite bias, but it still failed quicker than capacitors made with gold-platinum electrodes. This matter is still not resolved, but it is not likely that electrodes containing large amounts of silver will be reliable in screened capacitors unless hermetic sealing is employed. Of course, chip ceramic capacitors have been made with silver electrodes for years with no major migration problems, but the formats are quite different.

Glass Crossovers. These electrodes are wet by many glazes. Among the glasses which have been used to produce reliable crossovers are Corning 167JQ and 7052. A yellowish discoloration of such glasses tends to occur during firing due to diffusion of silver into the glass. Measurements of frequencies up to 150 mc with 167JQ glass in a capacitor geometry showed an open circuit however, implying no significant increase in conductivity of the glass. Among the other glasses which have been shown to wet these lands are Kimble TM7, Corning 7059, Drakenfeld E1313, 1527, etc. The fluxed electrodes do not tin well when fired on some of these glasses - probably due to interaction of the underlying glass with the frit in the electrode, which moves to the surface of the land and prevents tinning - but unfluxed lands and lands fluxed with lower softening lead borosilicate glasses have been shown to tin. Thus reliable glass crossover structures can be fabricated with fluxed silver palladium lands underneath, thick layers of glass as the insulation, and unfluxed silver palladium lands for the top tinnable region.

Tinnability. Properly formulated electrodes tin exceptionally well in lead-tin solders. This has been observed repeatedly both in the laboratory and in large scale tests. In any electrode system, optimized tinning conditions must be found by experimentation to fit the particular module fabrication process and production equipment. As with any other tinnable materials such as gold or copper, silver will dissolve in lead-tin solders, this erosion becoming more severe with increasing tin content, amount of agitation, time of immersion, and higher temperature. However for modules similar to SLT modules, this should not be a problem with 10 tin:90 lead solder at about $625°F$ for a reasonably long immersion time. For example, little or no noticeable erosion occurred on 30 second immersion or 3 consecutive 5 second immersions with such electrodes. Excellent tinning is also achieved in vibration or wave soldering equipment. As mentioned before, the proportion of erosion is strongly dependent on the amount of land present; thinner lands erode much more noticeably.

TABLE 9. Adhesion of Ag:Pd with Several Tinning Cycles.

Time in 10/90 Solder	Ag:Pd, 2% E1313			Ag:Pd, 2% 1527 4% Bi_2O_3		
	Low	High	Average	Low	High	Average
10 seconds	62	105	84	92	125	110
20 seconds	38	84	60	91	142	112
30 seconds	34	70	62	102	122	113

It would not be surprising if the wetting of silver palladium alloy is greater than gold-platinum alloy by solder. It has been mentioned that solder thicknesses are lower for silver palladium than for gold-platinum. This solder profile can be increased somewhat, but at the possible cost of a lower tinning yield if the solder is used at lower temperatures or the modules are withdrawn more rapidly from the solder. The thinner solder layer has not shown to be a particular problem, but of course, sufficient solder must be present to join active devices if desired.

Corrosion of these electrodes does not appear to be a problem; oxide and sulphide formation, if they occur on bare lands, are apparently removed in tinning - as evidenced by the ready tinnability of fired samples aged for as much as a year. One the lands are tinned, the corrosion of the solder is the pertinent problem. Tinned silver palladium electrodes have not shown as much corrosion under stresses of high humidity and temperature as tinned gold-platinum electrodes. As further evidence of the relative inertness of these lands, bare fired spirals 10 inches long and 5 mils wide were aged in air at $150°C$ for 900 hours. Although the lands darkened, the electrical resistance change was almost unmeasurable. There is evidence in the literature that palladium inhibits the formation of silver sulphide.[35]

Cost. Since normal organic vehicle systems have specific gravities of about 1, the volume ratio of metals increases in a startling manner as the weight ratio approaches high loadings. For example, each extra 1% by weight of vehicle adds 1cc. to 100 grams of formula, while each extra 1% by weight of gold adds only 1/21cc; 75% by weight of silver:palladium is about 21% by volume while 80% by weight of gold platinum is about 16.6% by volume. Not only is this effect evident in certain aspects of electrode performance (as conductivity), but also in cost factors. Precious metals are sold on a weight basis and lower density metals contain more volume per unit weight. Thus, silver and palladium show even greater cost advantages by volume compared to gold and platinum than by weight, and the amount of paste used per module is also a volume factor. Table 10 compares approximate cost factors at the market values existent in November, 1964. Recent market trends indicate even wider price spreads in the future. The Ag:Pd formula might be almost an order of magnitude cheaper than Au:Pt.

TABLE 10. Cost Factors.

Metal	Powder Cost/oz.	Powder Cost per cc.
Silver	$ 1.55	$ 0.55
Gold	52.00	35.00
Palladium	45.00	16.00
Platinum	120.00	81.60
80 Ag:20 Pd 69% Metal		
Au : Pt		

Reliability. Reliability of a fired conductor system can be expressed in many ways: the relative stability of properties during normal usage, such as strength, conductance, and other measurable properties, the absence of irreversible progressions of metallurgical changes such as intermetallic formation; and the relative stability to certain excessively high stresses. As previously mentioned, the electrical stability of both resistors and conductors are quite good. It has also been observed that certain strength characteristics degrade upon high temperature aging. The last two phenomena to be discussed rather briefly here are the aspects of intermetallic formation and silver migration under conditions of high stress.

Intermetallics. Since there is no gold present in this system, the typical brittle intermetallics that gold forms with such materials as copper and tin are not apparent. Although a tin-palladium intermetallic does form, this has not been shown to be detrimental in terms of strength characteristics. Minimizing brittle intermetallics very often shows an improvement on active device stability; in any metallurgical system, the amount of intermetallics generally is reduced by performing heating operations at the lowest possible temperature for the shortest possible time.

Silver Migration.[36-40] Silver is one of those metals which can form dendritic growths under conditions of high humidity and high field. Experimentation has shown that mechanistically the migration can be broken down into at least three components: dissolution of silver ions at the anode, transport across a very thin layer of moisture by means of ionic conduction (which may also involve a series of chemical precipitations based on solubility product), and finally plating of the silver on the cathode. There is considerable confusion about this mechanism, including the observation that migration can occur at temperatures sufficiently high to preclude large amounts of condensed moisture on the surface. It has been demonstrated that certain materials and coverings inhibit silver migration when alloyed with silver. These include magnesium and copper. A similar phenomenon exists with silver:palladium alloys also; the alloys migrate significantly slower than pure silver. Stress tests under water droplets have shown the alloy to be as much as an order of magnitude slower in migrating across a very small gap than pure silver.

However the important factor to be considered here is reliability under normal usage conditions. For example, it is conceivable that mechanistic changes are induced by severe stress tests. Experiments have shown, for example, that when migration tests are run in water-saturated atmospheres, many metals have been shown to migrate, including silver, copper, gold, bismuth, lead from frits, platinum, palladium, etc. Rather extensive test programs, under such conditions as $85^{\circ}C$, 85% relative humidity and fields of about 2 - 5 volts per mil, have indicated that silver migration from this alloy is no more severe than from the lead-tin solder. In the context of modules constructed and used in a manner similar to IBM's SLT process, silver migration is not a field problem under normal usage conditions. However, such a conclusion is applicable with some limitations; as electrical fields, substrates, surface conditions, ambient conditions, and fabrication techniques are changed for other products, migration tendencies should be assessed for those new conditions.

In summary, a compilation of the properties of this material compared with a commercial Au:Pt paste is presented in Table 11.

TABLE 11. Electrode Comparison.

Characteristic	Value	
	Ag:Pd	Au:Pt (DuPont 7553)
1. Flow Control	In formula	May require thickening
2. Porosity	Very dense	Somewhat porous
3. Adhesion-Tensile	$\cong$ 2300-4000 psi	$\cong$ 2500-4000 psi
4. Compression Resistance	$\cong$ 150K psi	$\cong$ 85K psi
5. Chip Strength (3 pads)	$\cong$ 300 gms	$\cong$ 240 gms
6. Flow Change, Humidity	None	Observed
7. Conductivity (untinned)	6-10X greater	
8. Tinned TCR	7-11%	18-19%
9. Resistor Drift	Equivalent	Slightly better on DuPont 7826
10. Resistor TCR	Equivalent	
11. Resistor Cracks		No electrical difference
12. Resistor Contact Res.	Better	
13. Interface Hotspot	Better	
14. Intermetallic Problems	Better	
15. Tinning Yield	Better	
16. Corrosion	Better	
17. Migration		Better
18. Volatility	Not Volatile	Volatile
19. In-House Resistor Compatibility		Incompatible
20. Settling in Jar	No settling	Settling
21. Paste Consumption	$\cong$ 30% less	
22. Cost	< $10/oz.	$\cong$ $60/oz.
23. Resistor Scaling		Better

Chapter 4

TERNARY ALLOY ELECTRODES

It is difficult to draw generalities to the properties of materials unless the processing parameters and the end goals are carefully defined. Since circuit manufacturers fabricate their modules in different ways, it is erroneous to assume that there can be a universally acceptable electrode paste system. As we have just seen, continued development and improvement of the more common binary alloys employed in circuit manufacturing have brought some of these formulations to a very acceptable level for this context. The most common material of this type has been gold-platinum, but it is expected that silver-palladium or gold-palladium[41] combinations will ultimately become more universally used.

Each of the two ternary alloy pastes described here represents a different interesting problem and its solution, providing illumination of the simpler binary alloy systems. The first ternary alloy - Ag:Au:Pd-represents an electrode material from which silver will not migrate and which is compatible with glazed doped resistors.[32] This ternary alloy has the concomitant advantage of lower cost than gold-platinum. The second alloy - Au : Pt : Pd[41] - addresses the difficulty of reproducibility of the gold-platinum system. Many properties of electrodes are dependent of the morphology of the particles in the paste, as well as the rheological characteristics resulting from the vehicle components; this alloy removes some of the raw material difficulties, achieves significant improvements in the resistance to erosion by solder, improves the fired density of the lands, and improves the adherence of the electrode to the ceramic substrates. It is indeed a usable electrode material, and performs better than some gold-platinum formulations in certain respects.

This chapter is divided into three main sections: an examination of some aspects of gold-platinum formulations (which will provide a background for the further development of the ternary alloys); and an intensive discussion of the formulation aspects of each of the two different ternary alloys. Finally, a comparison of various electrode systems will be made so that they can all be viewed at the same time, in the same context.

Unless otherwise stated, the processing details involved in this experimentation were the same as in the previous chapter.

GOLD-PLATINUM

Gold-platinum formulations have generally involved the ratio of either 3 or 4 parts of gold to 1 of platinum, combined with a flux predominately composed of bismuth oxide (with a possibility of a small amount of a borosilicate glass). A common difficulty with formulations of this type is the presence of fissuring, or cracking of the fired land. This fissuring has experimentally been shown to be a function of the flux, solids loading, pigment morphology, and vehicle, which will be discussed here.

Both the platinum and gold powder have rather strong influences on the tendency of the fired lands to fissure. However, extensive experimentation has been unable to predict fissuring with such powder properties as surface area or apparent density. Furthermore, inadequate correlation was found relating differences in particle shape or agglomerate structure, as viewed in electron micrographs. Heat treating the platinum in an inert or reducing atmosphere surprisingly did not correct the tendency of the alloy to fissure, even though this procedure is effective with platinum pastes. Large or dense platinum particles should sinter to less of a volume change than smaller or spongier particles. Indeed, both gray platinum and densified platinum did produce reasonably dense Au:Pt lands. But, — and this is one of those annoying balances that consistently reoccur in formulation work — since the particle-to-particle volume and dispersion relationships of the gold and the platinum were thus changed, more erosion in solder was noted in formulations of this type. This is attributed to the presence of more free gold, and should show up in x-ray analysis of lattice spacings of the fired alloy. The erosion can of course, be reduced by increasing the platinum-to-gold ratio, but this greatly increases the cost of an already expensive material.

The gold powder is also a factor in fissuring and land porosity. However, the choice of gold particles is rather restricted since large particles tend to flake out during the milling process of the paste. Thus, the malleability of the gold tends to limit formulations to very small particles of gold which, because of the large reductions in volume during sintering, adds to the fissuring and cracking problems.

Furthermore, due to the lowering of surface energies, large particles tend to produce lower adhesion values than similar formulations made with particles of higher surface area. In addition, larger particles tend to provide more secondary flow, since they do not cause gelled structures as do finely divided particles. As will be discussed later in the ternary alloy sections, the fissuring of the lands is also a function of the solids loading. In general, solids loading of about 78% or less by weight is often an acceptable limit.

The following experiment demonstrates the influence of changes in the glassy frit. When Drakenfeld E1313 lead borosilicate glass was substituted for the Drakenfeld 1527 glass (Table 1) used in a flux composed of a ratio of 1 glass to 2 bismuth oxide, fissuring became worse, and erosion of the land in solder was increased. Removing the glass completely and using only bismith oxide in the same percentage as the combination glass-bismuth oxide flux, did indeed provide a much denser land, but also extensive erosion. Obviously, the glass strongly controls erosion — perhaps by either providing a shield between some of the land and solder, or by holding the particles together during soldering. The bismuth oxide alone may be reduced by the rosin flux, or even by chemical reaction with the solder, whereas the new frit created by the interaction of bismuth oxide and some glasses is not removed. The softening point of such mixtures is distinctly different than pure Bi_2O_3 (Table 2).

Table 1. Glass formulae and properties.

	Coefficient of expansion 0–300°C x 10^{-7}	Softening point, °C	SiO_2	B_2O_3	PbO	Na_2O	K_2O	Al_2O_3	Other
E1313	70	517	22.4	8.8	66.6			2.2	
1527	63	588	30	9	51	1.7		3.2	5.1
210	114	390	10.4	2.9	84.7			2	
7052	46	710	65.5	24.1		4.1	4.1	2.2	
Lead bisilicate			32.8		64.8			2.2	

Table 2. Properties of mixtures of glass and Bi_2O_3.

Bi_2O_3	Glass	Coefficient of expansion 0–300°C x 10^{-7}	Softening point, °C
100	0		M.P. 820
60	40 1527	95.4	475
50	50 1527	88.3	501
30	25 210 45 lead bisilicate	90.3	497
30	25 E1313 45 lead bisilicate	82.6	527
	50 E1313 50 210	88.5	455
	E1313	70	517
	210	114	390

The vehicle, wetting agents, dispersion, and such processing parameters as drying and firing cycles, as well as land thickness, are also important in controlling the land structure. Commercially available gold-platinum pastes with relatively high volatility vehicles tend to fissure more severely when dried rapidly. In addition, the rapid drying can lead to considerably lower adhesion than that obtained by slower drying at low temperatures. Interestingly, such additives as terephthalic acid and ammonium sulphate control the secondary flow of the screened image and tend to densify the land with little sacrifice to other properties.[42] Obviously, the particles in the image should be allowed to settle after screening to form a dense compacted structure, or fissuring tends t be aggravated. This also supports the recommendation of a slow drying cycle.

Thermal gravimetric analysis of a commercial gold-platinum paste showed that oxidative degeneration is involved in the pyrolysis of the ethyl cellulose vehicle. It is not inconceivable that in some furnace atmospheres the volatilized pyrolysis products hover over the lands, forming either a neutral or reducing atmosphere in the immediate vicinity of the lands. This might lead not only to internal resin residues (the so-called "black heart" of the industry) in the lands due to incomplete pyrolysis, but also to possible reduction processes such as the conversion of the Bi_2O_3 to bismuth metal during firing. This latter reaction was suspected with platinum lands which were fluxed with Bi_2O_3. The obvious inference here — which is a well known axiom — is to have sufficient flow of fresh air through the furnace to wash away all pyrolyzed products.

Several simplified differential thermal analysis curves are shown in Fig. 1. No sharp phase transitions are discernible with the mixed metal powders from an Au:Pt paste, but if the upward curvature is a real one, it may imply alloying in the temperature range of about $600°C$ to $750°C$. As shown in curve B, the commercial gold-platinum does provide an exotherm at about $640°C$, which may either be diffusion into the 96% alumina substrate or reaction of the bismuth oxide and cadmium borosilicate glass in that paste. This does not necessarily imply that this is where the adhesion occurs; more complex mechanisms of sintering, reaction with the substrate glass, and viscous flow of the glass frit are also important.

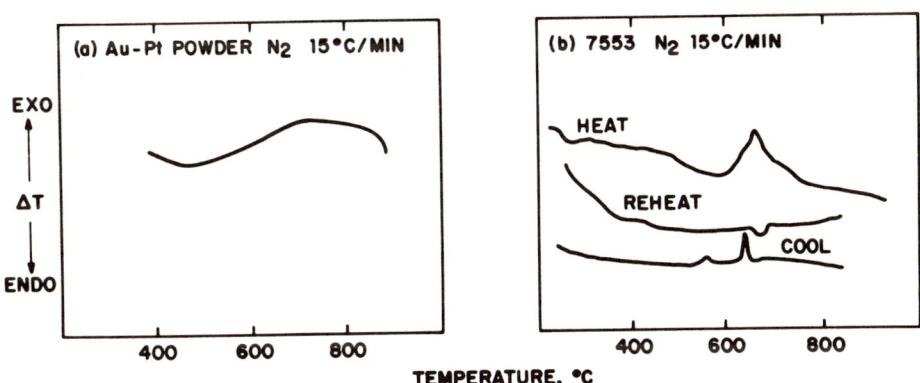

Fig. 1. Differential thermal analysis (simplified diagram) for Au-Pt paste.

We have seen several examples of how the properties of the ingredients in gold-platinum electrode paste affect the functioning of that paste. Many aspects have been omitted — such as the tremendous importance of the ratios of the metals to each other and to the glass frit. For example, the balance of wetting by solder against erosion by the solder is a fine balance indeed. With these complexities in mind, we now proceed to modifications of gold-platinum with a third added metal.

SILVER BEARING TERNARY ALLOY

Many module construction schemes require the use of glazed resistors. Although there is presently a large number of such glazed resistor systems commercially, we will discuss overcoming the high contact resistance of gold-platinum electrodes with doped glazed resistors.[32] These resistors — which are made with silver, palladium oxide, and a glass frit — are doped (as one would modify a semiconductor) to achieve stable low resistance values. The desired resistance range was in the order of about ten to several hundred ohms per square. Early experimentation showed that gold in Au:Pt electrodes acted as a strong contaminant at the interface of this type of resistor, and raised the contact resistance to a value sufficiently large to preclude its use. To overcome that contact resistance, small amounts of various metals and oxides were added to gold-platinum electrode formulations. The test format involved a pattern containing three resistors — which, upon perfect scaling (that is, little or no contact resistance), would provide resistances in a ratio of 1:2:3 (Fig. 2). As Table 3 shows, a platinum electrode fluxed with Bi_2O_3 scaled well with this resistor, and a silver-palladium electrode was also adequate, although lower resistance values were achieved from the smaller resistors than should have been expected. However, unmodified gold-platinum showed a high contact resistance and poor scaling indeed. As gold in the formulation was replaced with silver, gradual improvement in the resistance resulted, until finally, at about 20 parts of silver the resistance values matched the desired scaling. Similarly, additions of palladium made considerable improvement.

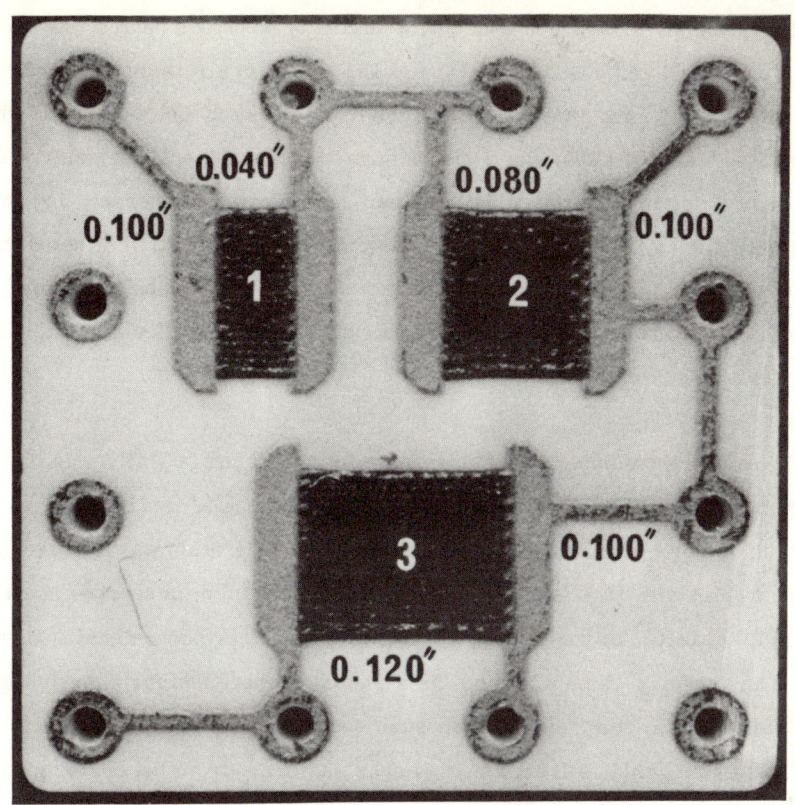

Fig. 2. Resistor test pattern.

Table 3. Scaling and temperature coefficient of resistance of low ohm glazed resistors on different electrodes.

Metal Flux or Addition	Pt, Bi_2O_3	80 Ag:20 Pd glass	Commercial Au:Pt*	In-house Au:Pt Bi_2O_3, glass	80Au:20Pd	*+5% Pd	*+5% Ag	PtAuAg 207010	PtA 20(
Scaling ratio									
Desired value 1	0.95	0.73	1.55	1.7	1.95	1.34	1.76	1.46	1
Desired value 2	1.92	1.85	2.3	2.5	2.2	2.15	2.57	2.13	2
Desired value 3	3	3	3	3	3	3	3	3	3
TCR %ΔR, 25°C to 75°	−0.6	−0.6	−0.2	0	−0.1	−0.3	−	−0.4	−0
TCR %ΔR, 25°C to 100°	−0.4	−0.5	−0.1	0.2	+0.1	−0.2	+0.1	−0.2	−0
TCR %ΔR, 25°C to 150°	0.9	0.9	1.1	1.3	1.5	+1.1	+0.6	+1.0	+0

Using the silver addition concept, a series of electrode pastes was prepared using a binder frit composed of Bi_2O_3 and 1527 glass in a ratio of 2:1, and replacing gold in 5% increments with silver from a base 80 gold - 20 platinum formulation. None of these electrodes migrated under a droplet of water with an impressed voltage; all tinned well without erosion, were adherent, and, at worst, only slightly fissured. As previously noted, 20% silver eliminated the contact resistance, and it was found that palladium could be substituted for platinum with a substantial cost saving without reducing the effectiveness of the electrodes.

To examine the upper range of this system, 25% and 30% silver in the gold:palladium:silver system were also tested. Both of these formulations eroded slightly in solder — the 25% silver just barely. No migration was observed under a water droplet with 40 volts impressed across fingers 5 mils apart — a very high field. In fact, this resistance to migration is rather phenomenal. In such tests, many species which would not be expected to move and form dendrites actually do so from other electrodes: such species as bismuth, lead (from the glass), platinum, gold, silver, tin, and even cadmium (from the cadmium plated clips used for the test) have been observed to migrate. Thus, it is particularly surprising that under similar tests, this ternary alloy did not migrate, indicating that all the materials are tied up tightly in an unexpected manner. As long as the silver is kept at or below the 25% level, this seems to be as inert an electrode system as could be desired. It is compatible with commercially available Ag:PdO glazed resistors, tins well, is adherent, does not erode in solder for 10 seconds, is somewhat more conductive than gold-platinum, and costs about one half as much. In addition, it does not require platinum — which is considerably more expensive than palladium and less readily available.

A series of experiments were performed to optimize this type of formulation. These experiments included studies of the flux, vehicle, and processing parameters. As previously mentioned, the original experiments were performed with electrodes fluxed with 2 parts of bismuth oxide to 1 part of 1527 glass. This produced sharp dense prints, but to improve the adhesion of the electrode, other frit materials were added to produce ternary combinations of glassy materials.

The first empirically derived adequate triple combination comprised 45 lead bisilicate:30 Bi_2O_3:25 210 borosilicate glass. A series of pastes with varying percentages of this triple flux was examined to combine the best tinnability with the least sacrifice of adhesion (Fig. 3). The adhesion essentially increases with percentage flux. However, the solderability — as might be expected — is worse with increasing flux concentration, as is the land porosity. As previously observed, erosion of the lands in molten solder became worse with lower flux percentages. The best balance of adhesion, solderability, erosion and porosity for this triple flux occurred between 8% and 9% flux.

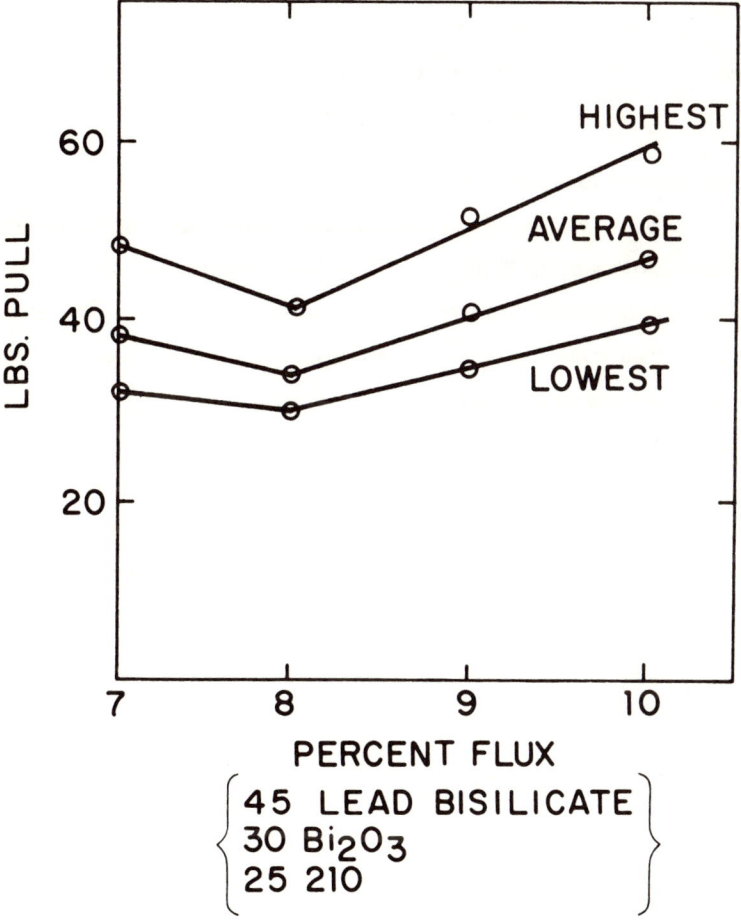

Fig. 3. The effect of glass flux on adhesion of Ag:Au:Pd.

Substitution of 1527 lead borosilicate glass for the 210 glass in this triple combination made a slight improvement in adhesion, but the lands fissured much more severely. On the other hand, substitution of E1313 lead borosilicate glass did not appreciably change the adhesion, but produced a denser land which tinned better. This last flux gave the best observed compromise of properties. As mentioned before with gold-platinum pastes, 78% solids loading produced a denser print and better tinning than 80% solids — perhaps due to better pigment settling in the wet print.

As shown in Table 4, the firing range of 835 to 900°C produces acceptable adhesion. Refiring the electrodes at 750°C (as might be done for glazed resistors) lowers the adhesion a very small amount. The recommended firing temperature for the best compromise of properties is about 835°C. Unlike gold-platinum electrodes, the drying cycle does not seem to be critical.

In summation of this ternary system, the optimum electrode contains the metal ratio of 55 gold:20 silver:25 palladium, combined with 8.5% of the flux ratio of 45 lead bisilicate: 30 Bi_2O_3; 25 E1313 lead borosilicate glass. Although this electrode functions well, its tinnability, conductivity, and cost are not as good as an optimized silver-palladium electrode, but somewhat better than gold-platinum. Figure 4 shows a ternary alloy diagram of Ag:Au:Pd[43] and Fig. 5 shows some properties of Ag:Au:Pt.[44]

Table 4. The effect of firing on the adhesion of silver:gold:palladium.

% Flux	Metals Au:Ag:Pd	Drying	Firing	Low	Adhesion Pounds Pull High	Avg.	Avg. psi
10	49:19:22	100°C - 15min.	835°C	42	49	48	2727
"	"	"	755°C	47	59	52	2955
"	55:20:25	"	835°C	41	54	46	2614
"	"	"	755°C	42	54	47	2670
"	55:20:25	"	835°C	57	68	62	3525
"	"	15hrs ~ 100°C	835°C	57	69	63	3580
8.5	"	100°C — 15min.	835°C 2-30-2*	40	49	44	2500
"	"	"	835°C 2-30-2* +755°C	31	42	37	2102
"	"	"	755°C	29	44	39	2216
"	"	"	900°C 15min.	33	46	43	2443
"	"	"	950°C	18	51	33	1875

*Firing cycle—minutes in, soak, out; other cycles - belt speed 1 1/2 inches per minute.

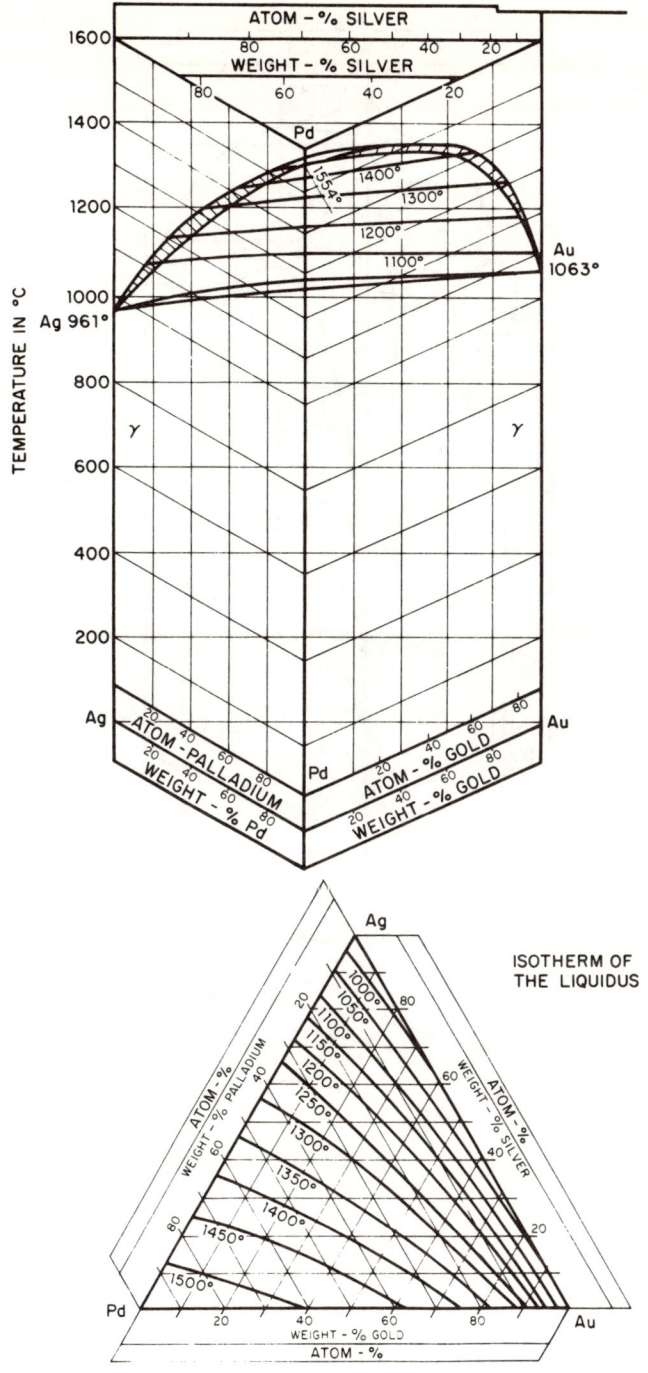

Fig. 4. Ternary diagrams of Au:Ag:Pd system.[43]

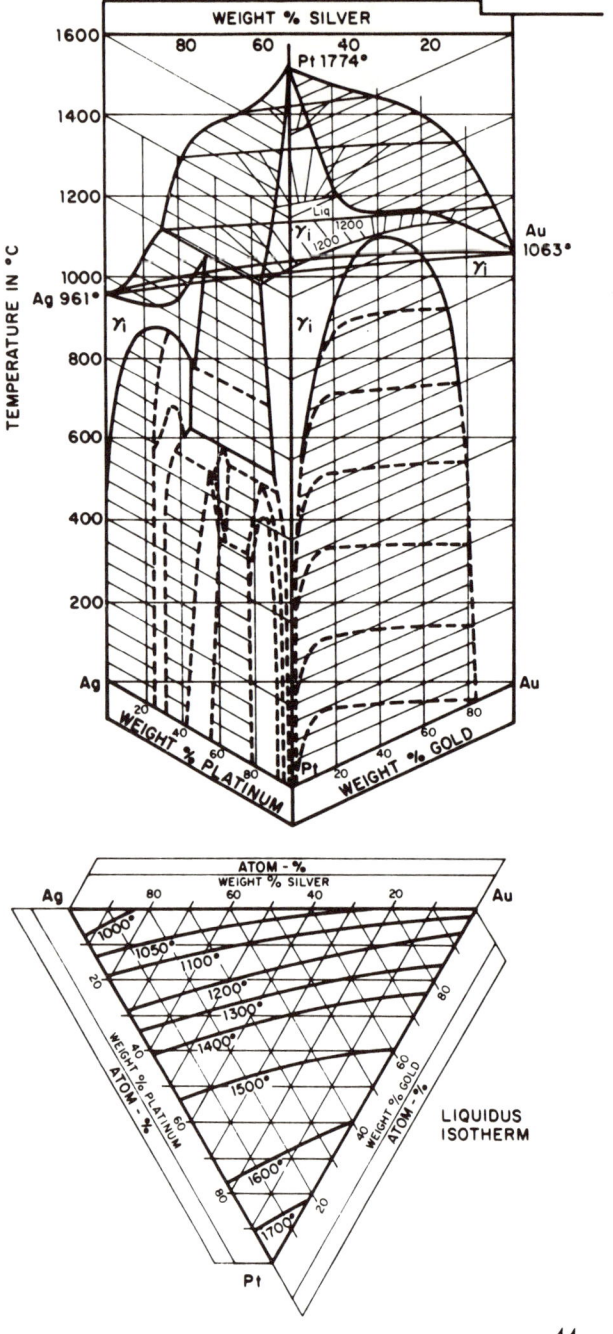

Fig. 5. Ternary diagrams of Au:Ag:Pt system.[44]

Ag	Au	Pt	Beginning of crystallization	Peritectic reaction		End of solidification
				Beginning	End	
10	30	60	—	1185	1153	—
10	40	50	—	1195	—	—
10	50	40	—	1158	—	1065
10	60	30	—	1200	—	1060
10	70	20	—	1230	—	1085
10	80	10	1178	—	—	1060
20	20	60	—	1170	1153	—
20	30	50	—	1145	—	—
20	40	40	—	1175	—	—
20	70	10	1155	—	—	1045
30	10	60	—	1170	1145	—
30	20	50	—	1165	—	—
30	30	40	—	1155	—	1040
30	40	30	—	1180	—	1065
30	50	20	1175	—	—	1032
30	60	10	1147	—	—	1040

GOLD:PLATINUM:PALLADIUM

We have seen how formulation changes can be made to modify the properties of gold-platinum electrodes. Replacing gold with some silver, and ultimately exchanging palladium for platinum, produced an electrode with an acceptable contact resistance for low ohm doped glazed resistors, at an appreciable cost saving. This section describes a different approach — that of improving the denseness of the lands and thereby attaining a high inertness to the erosion properties of solder, and higher adhesion. As previously noted, palladium can be added to gold-platinum formulae. If low surface area palladium (in the order of about 3 square meters per gram) is used substitutionally for gold, the land denseness is markedly improved without major sacrifice of other properties. Formulae with an initial gold-platinum weight ratio of 80 gold:20 platinum were prepared with successive replacements of gold by palladium in 5% increments to a maximum of 20% palladium. The denseness increased with each increment of palladium, without adverse changes in tinnability. In fact, the ratio 60 gold:20 platinum:20 palladium tins unusually smoothly. In addition, this ternary alloy seems to reduce the dependence of the fissuring on the platinum source — which was one of the goals of the formulation. To explore the higher range of palladium substituted for platinum in this system, formulations were prepared with the ratios 60 gold:15 platinum:25 palladium and 60 gold:10 platinum:30 palladium. Both were dense and tinned well, but the latter was a bit brittle after tinning (less adherent). Adhesion of the two pastes was lower than the 20% platinum paste. Thus, balancing cost and performance criteria, the 60 gold:20 platinum: 20 palladium ratio seems close to optimum, although 15% platinum may be acceptable.

These lands are particularly resistant to visible erosion in molten solder. No erosion was observed, for example, in six 10 second dips, or two dips totalling 1-1/2 minutes in molten 90% lead:10% tin at 625°F. No leaching was observed after active devices were reflowed to these lands; no other land system containing either gold or silver has been observed to be so resistant to erosion in solder, yet very wettable by solder. Another advantage of this metallurgical system might be a reduction of intermetallic formations with gold, since there is less gold available. However, there is no significant cost advantage of this formula when compared against gold-platinum.

Paste reproducibility can obviously encompass a broad range of factors — including land density, paste screenability, tinnability, adhesion, rheology, conductance, etc. This formula has been reproducibile with many variations in raw material lots, has at least three months shelf life, and shows no major processing defects. Very dense, tinnable lands have consistently been obtained with many different lots of raw materials with widely varying properties. All screened equally well, except for certain lots of platinum with extremely high surface area which produced gelation. This indicates that the platinum source has been made much less critical for these factors by the addition of the low surface area palladium. Of course, paste rheology will vary somewhat as the particle sizes and the surface area of the powders are altered, but this is a phenomenon which can be adjusted. Certainly, the rheology can be adjusted to specialized requirements such as increasing gelation for depositing very thick lands, or for adaptations to special printing and screening equipment.

The best reproducibility and performance can be obtained from this type of formula when the dispersion has been optimized, particularly in the milling process. This is obtained with the most effective shearing, a large number of passes, and cooled mill rolls. The better dispersions are generally superior in terms of shelf life, conductivity, and screenability. The vehicles used for this system involved low volatility butyl carbitol acetate, with ethyl cellulose as a thickening agent. To prevent secondary flow after the material has been screened, thixcin and MPA were used as gelation agents. These are hydrogenated castor oil derivatives, and it was found that when used as about 1% of the formulation, smooth screenable inks were obtainable. When different amounts of the MPA in the formulation were examined rheologically, it was difficult to distinguish the individual curves, despite the rather obvious differences in gelation. Casson plots (the square root of the shearing force versus the square root of the shearing strain), which should permit an estimation of yield value, showed superimposition of all these pastes with the same yield value. Thus, in order to truly characterize the gelation, experiments should be performed on the time rate of thixotropic recovery and rheological hysteresis measurements...which are difficult to experimentally reproduce accurately.

The amount of flux (again, 2 parts Bi_2O_3 to one of 1527) was varied, which again resulted in a monotonic increase in adhesion with flux concentration. This data is shown in Fig. 6. Even the highest flux concentration permitted smooth tinning. The higher flux concentration seemed to reduce some minor cracking at the edges of the fired lands. Surprisingly, there does seem to be an optimum land thickness; below a certain value, adhesion decreases. With 9% flux (which was estimated to be a good compromise), the data in Table 5 was obtained. Good adhesion has been shown to be consistently repeatable. Refiring the lands may even increase the adhesion a bit: 10% platinum alloy increased from 4054 psi average to 4148 psi (which may not be significant) after one refiring, and a 20%

alloy increased from 4545 psi to 4885 psi (which probably is significant) after three firings, and from 4375 psi to 4602 psi after one refiring. It was noted previously that low surface area palladium is desirable to minimize the fissuring aspects, and indeed, a higher surface area palladium (26 square meters per gram) gave only 3239 psi, probably due to fissuring effects (which is contradictory to the expectation of higher adhesion with higher surface energy). Palladium concentrations of 10%, 15% and 20%, gave over 4150 psi; this is unusually high adhesion — typical commercially available gold-platinum tends to run somewhere between 2500 and 4000 psi, with no assurance of reproducibility.

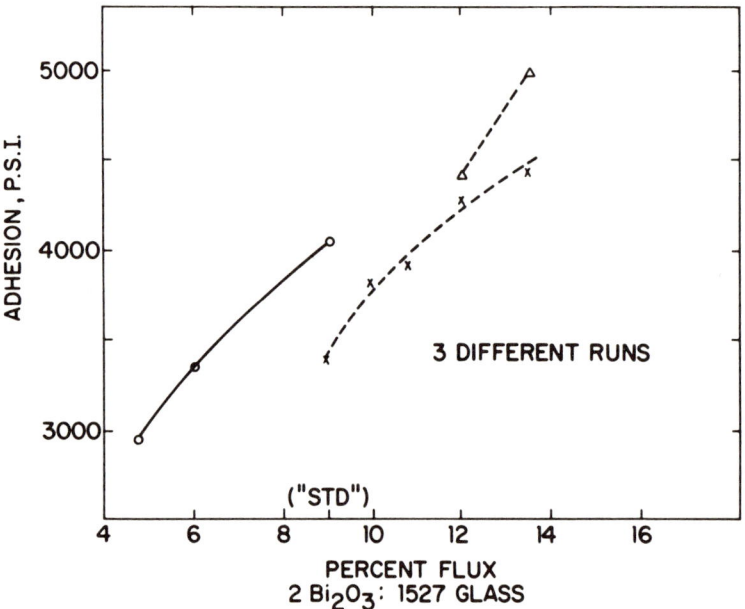

Fig. 6. The effect of glass flux on the adhesion of Au:Pt:Pd electrodes.

Table 5. The effect of land thickness on adhesion of gold: platinum: palladium electrodes. 9% - 2 Bi_2O_3/1527 glass.

Screen	Land Thickness, microns	Average Adhesion	
		Pounds pull	psi
open mask	8 - 10	59	3352
400 mesh	17 - 20	73	4148
325 mesh	18 - 22	69	3920
200 mesh	30 - 34	66	3750

As another indication of strength reliability, tinned samples of this ternary alloy were cycled from $0°C$ to $125°C$ for 522 cycles. Copper rivets were reflowed and pulled, as previously noted. Only a very small reduction in strength was apparent: from 4204 initial psi to 3693 psi; this is better than the greater degradation commonly observed under similar conditions with Au:Pt and Ag:Pd formulations.[30]

The refiring of the lands does not seem to disturb the tinnability of this ternary alloy system. However, when the lands are fired initially at temperatures higher than $750°$ (such as $835°$, $900°C$), some shrinkage, fissuring, and lifting can be observed. But when lands previously fired at $750°C$ were refired at $900°C$, they were still dense, adherent, and tinnable. Thus, if special module processing conditions require firing at temperatures above $750°C$, a prefiring at $750°C$ may be required to form these lands adequately on the ceramic substrate. No change in resistance was observed with increased milling, multiple firing, or refiring at a higher temperature (Table 6). However, a relatively small reduction in metal weight produces a rather large resistance increase. For example, 2.8% less metal by weight produced about 20 to 30% increase in land resistance. However, this particular experiment may have been exaggerated by the lower viscosity of the resulting paste, which probably deposited thinner lands than normal. The volume relationships of the materials in paste do not necessarily reflect the weight relationships, due to large differences in specific gravity. For example, ternary electrodes with the same metal volume as gold-platinum have a paste density of only 84% as much. Thus, for equivalent screened deposits, the ternary paste should be less costly (perhaps insignificantly so), but the tendency for secondary flow after screening due to gravitation forces (or the kinematic viscosity at very low rates of shear) should be better.

X-ray data shows incomplete gold:platinum:palladium alloy in firing from $750°C$ to $800°C$ (slowest cycle 1-1/4 inches/minute through the furnace). Although the ternary diagram of this alloy is not available, this is not unexpected; gold-platinum is not a continuous series of solid solutions, and the gold:silver:platinum diagram is quite complex (Figs. 4 and 5). The x-ray data reveals gold:platinum:palladium, plus free platinum and palladium.

Table 6. Effect of drying and firing on resistance of gold:platinum:palladium electrodes.

a. drying

		Resistance, ohms per inch					
		untinned			tinned		
	line width	5 mils	10 mils	15 mils	5 mils	10 mils	15 mils
100°C, 15 min.		31 - 65	9 - 10	5 - 5.5	5.5 - 9.5	1.5	.5 - .75
Avg.		45	9.5	5.4	6.0	1.5	.6
75°C, 60 min.		24 - 36	9.5 - 11.5	5.5 - 6.0	2.0 - 5.5	1.5	.5 - .75
Avg.		31	10.2	5.6	4.5	1.5	.55
Air dry 2 hrs. + 100°C, 15 min.		29 - 41	8 - 9	5.5	4.0 - 6.5	1.0 - 1.5	.5
Avg.		33	8.2	5.5	5.2	1.4	.5

b. Firing

		dried 100°C, 15 minutes					
		untinned			tinned		
	line width	5 mils	10 mils	15 mils	5 mils	10 mils	15 mils
725°C		36 - 50	11 - 12	5.5 - 6	5 - 5.5	1.5	.5
Avg.		39	11.2	5.5	5.2	1.5	.5
750°C		31 - 65	9 - 10	5 - 5.5	5.5 - 9.5	1.5	.5 - .75
Avg.		45	9.5	5.4	6.0	1.5	.6
775°C		35 - 43	8 - 8.5	4.5 - 5	6.5 - 7.0	1.5	.5
Avg.		39	8.1	4.9	6.6	1.5	.5
800°C		32 - 55	7.5 - 8.0	4.5 - 5.0	4.5 - 8.5	1.0 - 1.5	.5 - .75
Avg.		42	7.9	4.7	7.1	1.3	.55

The effect of drying and firing on adhesion and conductance was examined. Drying at 150°C produces less cracking along the edge of the lands than drying at 100°C, but care should be taken not to create blow holes or internal pinholes with excessively rapid drying. For formulations of this type, several hours of air drying followed by 150°C drying would probably be beneficial. In the firing range of 750°C to 800°C, no great difference was noted in adhesion, but drying is probably more important (though distinctly less so than with commercially available gold-platinum pastes), see Table 7. For adhesion, the best conditions of this series are 100°C drying and 750°C firing. More porosity is observed at 800°C and higher. With respect to conductivity, at higher firing temperatures the resistance of the tinned narrow 5 mil lines was somewhat higher, indicating a thinner solder layer on these lines. The best conductivity was obtained from the samples which were air dried for 2 hours before 100°C drying.

As one further aspect of the use of this material in a circuit module, compatibility with glazing material was examined. Corning 7052 glass was screened and fired on top of a previously fired 10 inch long, 5 mil wide spiral of the ternary lands which had been screened through 200 mesh. The fired lands had ranged from about 200 to 300 ohms, and the resistance decreased by about 2 ohms (or

Table 7. Effect of drying and firing on adhesion of gold: platinum palladium electrodes.

a. drying	fired at 750°C
100°C, 15 minutes	4602 psi average
2 hrs air dry + 100°C, 15 minutes	4100 psi average
75°C, 60 minutes	3920 psi average

b. firing	dried at 100°C, 15 minutes
725°C	4100 psi average
750°C	4602 psi average
775°C	4318 psi average
800°C	4715 psi average

(all firings 15 minute rise, 15 minute soak, 15 minute cool)

about 1%), probably due to further compaction of the lands. This indicates that resistance changes due to interactions with glass of this type are not a problem. The lands were dense under the glass — implying that for circuits utilizing these lands with glass crossovers, resistance ranges will be better controlled than with lands that tend to fissure more. This glass — 7052 — was also cofired with the lands. Again, no fissuring or obvious interactions were observed.

In summation of this ternary formulation, we see that when gold is replaced by palladium in a gold-platinum formula — and particularly, if the palladium has a low surface area — the lands can be made much denser and more reproducible. An unexpected increase in the resistance to erosion in solder is also observed. There is little cost differentiation, since gold and palladium powders tend to be of about the same value, although some small saving might be observed in the lower specific gravity of the palladium compared to the gold. Table 8 summarizes the effect of formula modifications in the Au:Pt:Pd system.

CONCLUSIONS

We have seen that what might be considered as insignificant details in the formulation of fired electrode pastes can be rather significant in the functioning of these pastes. Every aspect of each formulation has its areas of criticalness (and there are aspects not covered in this chapter involving, for example, the vehicles). To blithely discuss the properties of these materials without knowing the specifics of their formulation can be misleading. Thus, it does not make much sense to talk about the exact properties of gold-platinum electrodes or silver-palladium electrodes or even ternary electrodes in general, without defining the specific formulation and processing details. There are good formulations and bad formulations of each type. Furthermore, the definitions of good and bad certainly rely on the context. A manufacturer making a noncritical circuit might not be concerned about conductivity, while another manufacturer might find it absolutely necessary to have the very best conductivity for his circuits.

To expand on the comparison between gold-platinum and silver-palladium, Table 9 summarizes some properties of these materials and the two ternary alloys. All of these alloys should be reliable and useful under the proper conditions.

Table 8. Effect of formula changes on properties of Au:Pt:Pd electrodes.

	Ratio	Material	%	Preferred type	Effects of Changes	
					Increase concentration	Vary type
72.2% Metals	60	Gold	43.5	Hanovia #2 Brown gold	Increases conductance, erosion,	Large particles or flakes—plating out on mill, lumps.
	20	Platinum	14.6	Engelhard #4 Pt black	Increases cost. ex cess decreases adhesion.	Low surface area—less adhesion. High surface area—gelation, high viscosity.
	20	Palladium	14.6	2-6 m^2/gm. surface area	5-20% increases denseness. Excess decreases adhesion, tinnability.	High surface area—increases fissuring, gelati lowers adhesion.
7.3% Flux	2	Bi$_2$O$_3$	4.8		Increases erosion, tinnability	increase adhesion, improves edges, decrease tinnability	No data
	1	1527 glass	2.5		Decreases erosion, tinnability	
20% Vehicle		Thixcin gelation agent	0.2	Thixcin R	Increases gelation, clogging, mesh marks, edge effects	MPA—better edges, poor shelf life. Aluminum stearate —may disturb tinnability
	18	B.C.A. Butyl Carbitol Acetate	15.4		Decreases viscosity	No data
V149	17	Ethocel Ethyl Cellulose Resin	3.4	Fisher L.V. grade	Increases viscosity	Lower molecular weight improves edges.
	5	CO430 surfactant, lubricant	1.0	IGEPAL CO430	Decreases viscosity. Reduces clogging. Improves flow, edge effects	Sarkosyl O—improves edges, wets well, reduces viscosity, causes slightly more porosity.

Table 9. Summary of electrode properties.

	Ag:Pd^3	Au:Pt^3	Ag:Au:Pd	Au:Pt:Pd
Land denseness	good	potential problem	porous	*good
Adhesion-tensile psi	2300 - 4000	2500 - 4000	2000 - 3000	*3900 - 5000
Wetting by solder	*G	G	G	G
Erosion by solder	F	F - G	F - G	*G
Resistance (Ω/inch 200 mesh, .010 inch line)	*1.5	6.5 - 10	5	7 - 10
Interaction with glaze	G	G	G	G
Reproducibility	G	F	F - G	G
Density (gms/cc)	3.1	4.7	—	3.9
Cost/oz (appr.)	*≅$10	≅$60	≅$35	≅$60
Insensitivity to processing	G	P	F	G
Migration	F	G	G	G
Doped resistor compatibility	G	P	G	F
General reliability	G	G	G	G

*Outstanding property

Chapter 5

GLAZE RESISTORS

Introduction. Users of thick film resistors should be particularly aware that even minor and subtle paste changes can produce large fluctuations in the electrical properties; certainly, any changes in the raw materials themselves are likely to perturb them. Such changes might include variations in the ratios of the metals and glass powders or in the ratio of the solids to the liquid vehicle.[45] Other changes, as in particle sizes,[46] chemical purity, degree of oxidation,[47] surface activity, degree of particle aggregation, moisture content, etc. may less expectedly result in a lack of batch-to-batch uniformity.

Of greater concern to the paste user are the resistor processing details, since he can manipulate them more readily. These include such extensively discussed factors as: morphology of the screened deposit -- the paste rheology, and screening operations which effect it; drying of the print and firing conditions -- from both the standpoint of uniformity and the particle packing and degree of oxidation; interactions with the substrate; interactions with the conductors, etc.[45, 48] Indeed, several of these parameters will be mentioned again as this discussion proceeds, since they all interact -- sometimes better than expected, sometimes disastrously.

Two types of resistors are examined in this chapter: predominantly, the silver-palladium oxide-glass type,[49] although some examples of doped indium oxide glaze resistors are included to show that some of the effects are not restricted to the former.[33] The generic formulas and the identifications which will be used are shown in Table 1. The conductors were either 80 silver/20 palladium or 75 gold/25 platinum formulations, unless otherwise designated. They were commonly applied through 325 mesh or 200 mesh stainless steel screens, or through etched metal masks with approximately 1 mil cavity. Resistors were normally applied through 200 or 165 mesh stainless steel screens. A common drying cycle was $100°C$ for 15 minutes, and the samples were usually fired at $750°C$ for about one half hour. To avoid confusion, specific processing details are included with a description of each experiment. Ceramic substrates of 94 - 96% alumina were employed and sampling normally included about ten of each condition.

Table 1. Typical Formulation

Ag:PdO: glass type:

18 Silver
22 Palladium oxide   } 80% solids in vehicle
60 Borosilicate glass

(low resistance pastes were doped with Lithium as the carbonate)

In_2O_3 type:

1.4 Sb_2O_3
58.6 In_2O_3 } 80% solids in vehicle
40   Glass

Vehicle: 37 resin
             6 surfactant
             57 solvent

Conductor:

80 Ag:20 Pd
5 Glass frit binder   } 75% Solids
Vehicle   20%

<u>The Vehicle-Formula</u>. The liquid binder which carries the particles of metal and glass and forms a paste-like structure is commonly composed of at least three ingredients: A resinous or polymeric binder to provide the basic rheological properties, a solvent to dilute the resin (and which can be evaporated later to dry the image), and perhaps a surface active agent to permit the solid particles to be wet by the vehicle and properly dispersed in it. Beyond these rather obvious requirements are a series of more subtle limitations which are worthy of consideration in the choice of these materials. For example, metallic constituents of the vehicle might result in doping of the glaze resistor, and therefore it is wise to have as metal-free a vehicle as possible. For example, normal surfactants might contain some sodium or potassium, but these ions are known to migrate rapidly through glassy substances and affect drift phenomena. Fortunately, non-metallic vehicle constituents are available so that this potential problem can be readily avoided.

The volatility of the solvent system is also very pertinent, and it is one of the most important factors relating to the ease of screening. Excessively volatile solvents will tend to dry out during usage, and thus cause fluctuations in the obtained resistance values, since the amount deposited changes as the solvent leaves the paste. Using a solvent of very low volatility is therefore desirable, but this puts extra burdens on the formulator to control the flow of the print on the substrate and to provide good wetting of the pigment (metal and glass) particles; low volatility liquids tend to have decreased dissolving and wetting powers compared to smaller more volatile molecules.

<u>Emulsion Effects</u>. The polarity of the vehicle is an aspect which is significant to users of screens with photosensitive emulsions defining the circuit pattern. Hydrophilic solvents, which are highly polar in nature, will tend to swell and soften the emulsions -- thus seriously decreasing the usable lifetime as well as causing changes in the actual emulsion pattern dimensions. This is not a consideration when etched metal masks are used, but there is still rather widespread use of polyvinyl emulsions on screens.

To examine this aspect, strips of Screenstar 82 emulsion were drawn on a glass plate sprayed with a flourocarbon release agent. One inch wide strips, approximately 1 mil thick (using tape as guides) were pulled from the glass after the emulsion was dry. The strips were exposed to an arc light in the normal manner, and then immersed in various liquids for a known length of time.

Two series were run: the first for immersion over a weekend, after which the strips were removed from the liquids and washed with tetrachlor and dried; the second for immersion overnight, with no washing (the strips were blotted with absorbent paper and dried in normal ambient conditions for 3 hours). Several evaluations were made of the treated strips:

1. Weight change.
2. Optical density.
3. Appearance.
4. Tensile strength.

The data is summarized in Table 2. Weight change does not tell much; such a test may either be non-informative or even misleading -- perhaps due to experimental difficulties. For example, no weight change was observed with terpineol + ethyl cellulose, yet this liquid decreased the tensile strength considerably. Furthermore, a weight gain was noticed with tetrachloroethylene which is not explained.

Any loss in optical transmission should indicate some form of attack by the liquid, either of a chemical or solvation nature. Indeed, those liquids which obviously deteriorate film strength do reduce optical transmission drastically (as trichlor, terpineol etc.). Furthermore, an aliphatic ink oil 10-550[*] which did not reduce the film strength did not produce an appreciable increase in optical density. Thus, although this test is neither quantitative or precise, it does give interesting indications.

Few liquids drastically effected the film appearance, with notable exceptions. Trichlor gave a bleached appearance, while terpineol + ethyl cellulose gave a sticky feel. As might be expected, water and denatured alcohol softened the film beyond usability - it curled and stuck to itself. In general, the more polar the liquid, the more severe the attack.

---

[*] AMSCO

Table 2. Effect of Liquids on Screenstar 82 Emulsion Strips.

Liquid	Optical Density (% transmission)			Percent Weight Change	Appearance	Tensile Strength (% of 5 pounds)		
	Series 1	Series 2 Control Immersed				Series 1	Series 2 Control Immersed	
Terpineol + Ethyl cellulose	2.5			0.0		29.5		
Alpha Terpineol	3	8	4	-0.1		41	53 / 47	42 / 38.5
Beta Terpineol	4.5	7.5	3.5	-3.2		35	33.5 / 39	44.5 / 48
Dibutyl Phthalate	4	9	-7	-3.9		42	48.3 / 48	47.5 / 45.5
Butyl Carbitol Acetate	6	10	9.0	-20.0		34	44.5 / 49	46.5 / 46
Tetrachloroethylene	5			+8.8		40		
Trichloroethylene	1			-36.8	faded, brittle, opaque	28		
10-550 Amsco Ink Oil	9.5	7	6	-12.7		51	48.5	61.5
Dentatured Alcohol	--			-5.3	color, change, soft, tacky			
Tap Water	--				color, change, soft, tacky			
Untreated Emulsion	9.5-10.5					53.9		

Rheology. Several aspects of rheology are particularly important in these systems. Long shelf life can only be achieved if the system is virtually non-reacting; any unsaturation in the polymer or the presence of oxidizable groups are likely to react over a period of time, particularly in the presence of the active metal systems often employed in such resistors. Palladium and palladium oxide can be catalytic to oxidation of double bonds and decarboxylations. This need for non-reactivity further complicates the resin-solvent solubility aspects.

The resin molecular weight and configuration are also very important. Excessively long and kinked molecules will tend to "string out" when the screen is separated from the part being printed, and these strings can snap back to the module surface to cause contaminating debris. Alternatively, if the molecular weight is not sufficiently high, a large amount of resin is required to achieve the proper screening consistency. This can lead to erratic behavior since the loss of even a small amount of the solvent that remains will cause rather large increases in the viscosity. Thus, a compromise in molecular weight is called for, or a mixture of molecular weights.

The most important rheological characteristics are the viscosity at the shearing rate used during the screening operation, the viscosity at very low rates of shear (gravity) which are experienced by the print after screening, and the time rate of the thixotropic buildup -- or the ability to reform a gel structure and thus become re-immobilized. Many screening systems employ a gelation agent to provide such a gel structure, but in the formulations described here, this was found to be unnecessary. The high pigment concentration, its finely divided nature, and the inclusion of a small amount of colloidal silica are sufficient to produce a sufficiently gelled paste to preclude the use of additional gelation agents. If such an agent were required, the same consideration with respect to metal content would apply as to the other vehicle constituents. Therefore, metal-containing salts are not preferred. Fortunately, as with resins, solvents, and surfactants, there are a wide variety of usable gelation agents which do not contain metal ions. Although bentonites are not particularly desirable, hydrogenated castor oils work quite well. An interesting aspect to consider in the choice of these agents is the gel temperature, above which the gel properties are destroyed and fluidity is regained by the system. It is quite foolish to use a gel system and dry it above its gel temperature, since this destroys the effectiveness of the gel.

Rheological stability is related to atmospheric humidity; hydrophilic vehicles will normally change their viscosity characteristics as the moisture content of the environment changes.[50] This is less critical with hydrophobic vehicles, which gratifyingly ties in with the desire for non-polarity to avoid softening and swelling in the screen emulsions. (This is one aspect in the complex field of paste formulation which is not self-contradictory). However, the vehicle is not the only ingredient which can be affected by the humidity of the atmosphere. Hydrophilic pigment surfaces can also be effected, and it is known that the pigment-vehicle interface can be strongly effected by humidity.[57] With the proper use of surfactants, this vehicle-pigment interface can be adequately maintained and thus provide rheological stability throughout large changes in humidity. Non-ionic surfactants such as polyoxyethylene derivatives are quite useful (e.g., Igepal C 0430.* The more hydrophobic types of these materials are preferred.

---

*G.A.F. Chemical Co., Antara Div.

Commonly used resistor vehicles have contained such relatively polar resins as ethyl cellulose or methacrylates, diluted with relatively polar solvents (i.e., terpineol, carbitol, or cellosolve and their relatives and derivatives). Some manufacturers recommend that if further dilution is required, a less volatile solvent such as butyl carbitol acetate — which is somewhat polar, but less so - be used. However, to answer the above needs, non-polar hydrocarbon solvents and resins can also be utilized in such resistor pastes. Among such resins which have been found to be particularly useful are polystyrenes, polyterpenes, and poly alpha methylstyrene.[52] The latter resin is particularly interesting, since it has an unusually high rate of thermal degradation, and thus can be driven rapidly from the composition during the firing process. At $350^\circ C$, for example, it pyrolizes over a hundred times faster than polystyrene, and probably decomposes by an "unzipping" mechanism, so that a very high yield of monomer (greater than 95%) results.[53] This will leave less carbonaceous residue than polystyrene, which has an approximate yield of 40% monomer. As will soon be shown, the pyrolysis properties of the resin and solvent can indeed have an effect on the resistor properties.

Such resins are soluble in non-polar vehicles. However, pure substances which are both non-polar and have high solvency power (expressed as Kauri-Butanol values or Aniline values) are not very available, and the prices are relatively high. Fortunately, some of the high boiling petroleum distillate fractions have the desired solvent properties. Of particular interest are solvents with a high naphthenic aromatic hydrocarbon content. Such solvents are available with the preferred vapor pressures in the range of 10 to 20 mm. at $212^\circ F$, but in the range of 0.001 to 0.01 mm. at room temperature.[54] These ranges insure that the paste will be stable and nonvolatile at normal storage and working conditions, but will be driven out gently and with sufficient speed at the drying and firing conditions. Two lower polarity, low volatility solvents - Amsco' HCC solvent, and Union Carbide's butyl carbitol acetate - will be emphasized in this paper. Figures 1 and 2 provide some typical volatilization characteristics of a few pertinent solvents and vehicles. The room ambient curves were obtained by periodically weighing aluminum weighing cups of solvent which were placed in the flowing air of a chemical hood. The higher temperature curves were obtained from on Ohaus moisture balance calibrated at 85°C. Although very great volatility differences occur at room temperature, these differences are less significant at drying temperatures. Thus very low volatility solvents can be used in such formulations without interfering with the drying of the prints. Of course, solvent retention in the resin is an aspect not clarified by this sort of data.

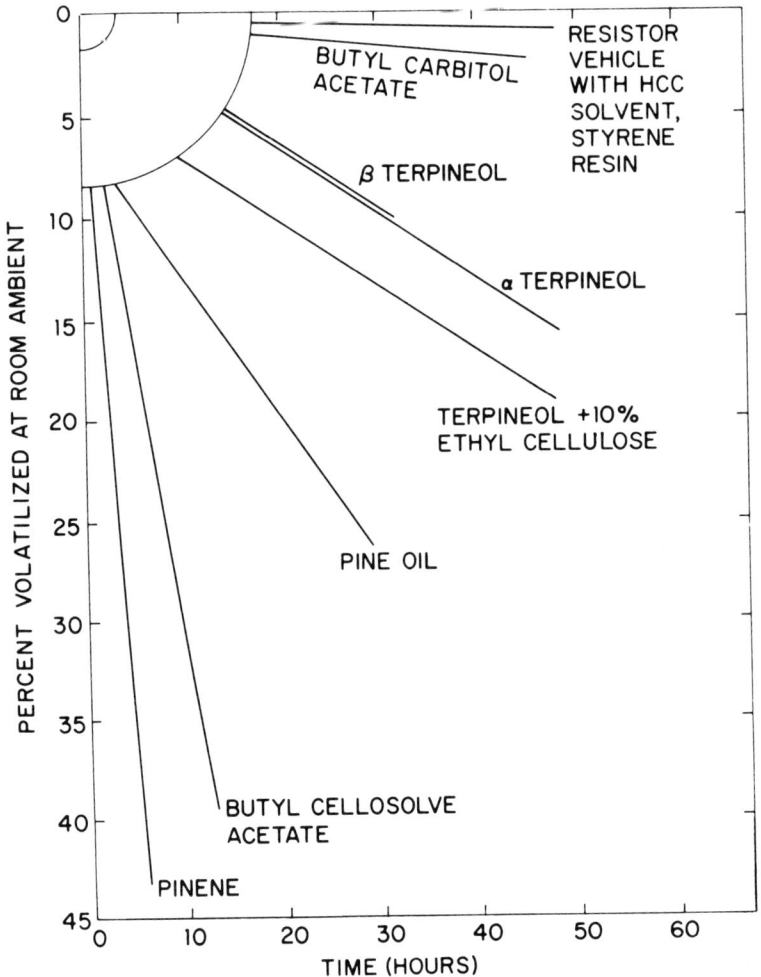

Fig. 1. Relative evaporation rates in room ambient flowing air.

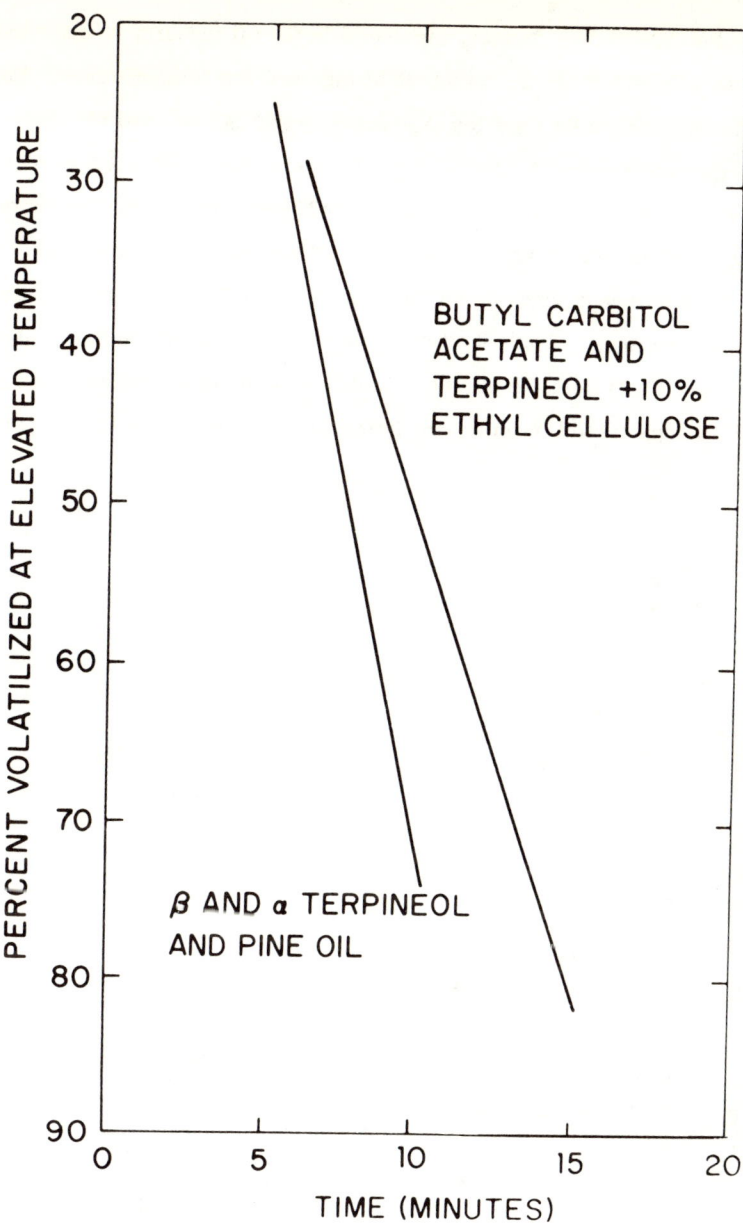

Fig. 2. Relative evaporation rates under radiant heat, approximately 85° C.

The viscosity coefficient of liquids normally lowers as the temperature increases. Of course, it is desirable for this reduction in viscosity to be minimal, so that less stringent ambient control during processing would be necessary. Certain types of liquids have highly irregular kinked and curled molecules with extremely low T.C.V.'s. In fact, these enable the oil industry to make multi-grade motor oils which can be used both at high and low temperatures. Certain polyisobutylenes are quite good for the purpose, and can be used for flattening out the temperature-viscosity curve (Fig. 3). However, an excess of such a material will result in the formation of a very "long" paste, which tends to string out during the screening operation, and thus cause undesirable residues. In addition, the polyisobutylenes do not burn out well, and are capable of forming carbonaceous residues. Table 3 summarizes some of the molecular configurations which affect both the elasticity and flow of polymer systems. Such interrelationships are often important in the design of solutions for proper rheology.

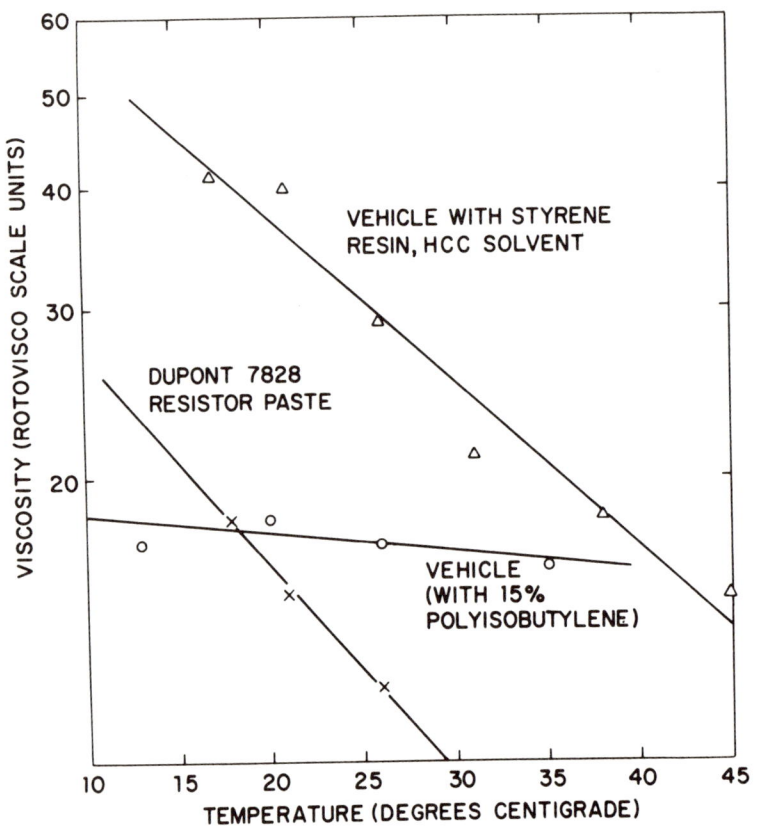

Fig. 3. Temperature-viscosity relationship for resistor pastes.

Table 3. Effect of Molecular Confuguration on Elasticity and Flow.

Factor		Elasticity	Flow
molecular size	increasing in size	I	D
configuration	kinkiness increasing	I	D
configuration	curliness increasing	I	I
configuration	side groups increasing	I	D
bonding	increase of single bonds	I	I
bonding	increase of intermolecular forces	I	I
bonding	polarity increasing	D	D
solvation	same type solvent	D	D
	smaller size solvent	I	I
	plasticizing increasing	I	N
shape	ring structure increase	D	I
cross linking	heavy	D	D
cross linking	moderate	I	D
steric hindrance	increase	V	D
crystallization	increase of crystallization	I	D
D - decrease I - increase V - varies N - no change			

This section has shown some of the complex relationships of the vehicle ingredients, particularly with respect to the process through screening. The next section will show a few examples of how these ingredients can also affect resistance properties.

Vehicle Effects on Resistor Properties. The first illustration of how the vehicle components can effect the resistance values involves the changing of the resin and solvent. Table 4 lists several groups of resistor formulations, the members of each group being identical with respect to everything except the vehicle. The resistance values are presented in ohms per square, for a resistor approximately 1.2 squares (0.100 x 0.120 inches). These resistors were deposited through 165 mesh screens and were approximately 52 microns thick after being dried. A series of seven different firing cycles are indicated, all with a peak soak temperature of $750^{\circ}C$ and a soak dwell of 13 minutes. The three numbers in each cycle represent the amount of time spent in coming up to temperature in the CM test furnace, the amount of time at that soak temperature, and then the amount required for the rack to come out of the furnace into the room ambient. In the first five cycles the <u>in</u> and <u>out</u> time lapses were equivalent, while the last two cycles shown non-symmetry. Cycle D of this series was chosen as a reference point, and the resistance values for that firing condition were chosen to represent 100%. Thus the resistance values obtained for the other firings are expressed as percentages of that value to permit more ready comparison.

Table 4. Effect of Resin on Resistance and TCR for Different Firing Conditions.

Resistor Compositions:

1. Ag:Pd Ratio - 0.9
   60% Borosilicate Glass
   Doped with 0.35 gm Li_2CO_3 in the PdO   } 80% Solids
   Remainder is vehicle, which contains:
   37% Resin (such as Poly Alpha Methyl Styrene)
   Remainder 3-6% Nonyl Phenoxy Polyoxyethylenethanol (IGEPAL C0430-surfactant); Aromatic Napthenic Solvent (AMSCO HCC Solvent)

2. 15.1 PdO Doped with 1% Li_2CO_3
   11.8 Ag
   37.5 Borosilicate Glass   } 80% Solids
   2 CABOSIL
   Remainder is vehicle, as above

3. Ag:Pd Ratio - 0.5
   48% Borosilicate Glass   } 79% Solids
   Doped with 2.5 Gms Li_2CO_3 in PdO
   Remainder is vehicle, as above

165 Mesh screen
Deposits ≃ 52μ, dried
1.2 □ resistors
Ag:Pd Conductors

	Firing Cycle (Soak 13 Minutes at 750°C) Resistance, ohms/□							T.C.R. (Temperature Coefficient of Resistance, 25°C - 100°C)					
	A	B	C	D	E	F	G	A	B	C	D	E	F	G
Rise Time (Min.) Fall Time	15 15	17.5 17.5	21 21	25.4 25.4	29 29	15 25.4	25.4 15							
(Resin)														
1a. 37% Styrene in HCC	80 (50%)	118 (74%)	142 (89%)	159 (100%)	174 (109%)			.7	.2	0	-.2	-.3		
1b. 37% Alpha Methyl Styrene (M.P. 290°C)	130 (84%)	132 (86%)	153 (99%)	154 (100%)	162 (105%)			-.3	-.3	-.5	-.6	-.6		
1c. 37% Alpha Methyl Styrene (M.P. 240°C)	162 (84%)	168 (87%)	181 (94%)	194 (100%)	203 (105%)			-.4	-.7	-.7	-.7	-.7		
2a. 37% Styrene in HCC	75 (51%)	99 (68%)	118 (81%)	146 (100%)	149 (102%)	80 (55%)	134 (92%)	.6	0	-.1	-.3	-.4	+.4	-.3
2b. 37% Alpha Methyl Styrene (M.P. 240°C)	137 (80%)	148 (86%)	154 (90%)	172 (100%)	170 (99%)	139 (81%)	161 (94%)	-.4	-.4	-.4	-.5	-.5	-.4	-.7
3a. 37% Styrene	169 (122%)	153 (111%)	150 (109%)	138 (100%)	139 (101%)			.4	0	-.3	-.5	-.6		
3b. 37% Alpha Methyl Styrene (M.P. 240°C)	76 (95%)	79 (99%)	75 (94%)	80 (100%)	81 (101%)	78 (98%)	79 (99%)	-.1	-.1	-.2	-.3	-.4	-.2	-.2

Two main attributes stand out from this data: not only does the actual resistance value change according to the vehicle used, but the affect of the firing cycle is also different. In example 1a, 1b and 1c, the vehicles using poly alpha methylstyrene dissolved in butyl carbitol acetate show a much lower dependence on the rise and fall time of the temperature than does the vehicle with styrene dissolved in HCC. This is consistent with the rapid pyrolysis behavior of the poly alpha methylstyrene resin mentioned previously. The resistance values themselves are higher, probably because there is less reducing action from the poly alpha methylstyrene when it is pyrolyzed, compared to the styrene.[53] It is axiomatic for resistors of this type that the atmosphere around the resistors during firing must be oxidizing in nature, and any reducing action will tend to change the ultimate resistance values.[48] Examples 2a and 2b show the same effect. This implies that resins which decompose more readily are preferable in formulations of this type. It also suggests a means for adjusting resistor values by varying the vehicle components in resistor formulations to compensate for the changes in properties of the pigmentation.

Table 4 also shows some interesting relationships of the temperature coefficient of resistance (T.C.R.) and the firing cycle. The T.C.R. varies more for the styrene vehicle than for the poly alpha methylstyrene vehicle. It is true that the latter T.C.R.'s were larger (in a negative direction), but they were less suspectable to changes in the firing cycle. Thus, the use of such a resin permits the formulation of the other raw materials to create a T.C.R. close to zero, and then fluctuations in the furnace will not be as significant a factor. On the other hand, with vehicles which do not burn out so easily and interfere with the oxidation-reduction mechanism of the resistor, the T.C.R. should be expected to vary considerably more as the furnace speed or firing conditions fluctuate.

The second illustration of these phenomena appears in a different format in Table 5. Here, a formulation was kept constant, except that the ampunt of styrene resin in the vehicle was varied. Of course, the less resin in the vehicle -- all other things being equal -- the lower the viscosity of the paste will be, and in general the thinner the deposit. Thus, an increase in resistance should be expected as the amount of resin is decreased, since the resulting deposit is thinner.

However, the T.C.R. obviously moved in a negative direction as the amount of resin decreased. There is no apparent monotonic change in the resistance values during storage at 300°C but the drift of the resistors under power is much greater in a negative direction for the lower resin - thinner resistors. From this data, it is difficult to determine whether the increased negative drift is due to resin affects or resistor thickness affects. I suspect the latter predominates here.

Table 5. Effect of Resin Concentration on Resistance Values and Drift.

	% Styrene in HCC	Viscosity[1]		Resistance ohms/□			Weight Deposited (GMS)	% ΔR[2] 300°C Storage	%ΔR[3] Power Test	TCR 25-100°C
		Vehicle	Paste	0.4□	0.8□	1.2□				
4a	39	22	26	579	1304	2320	0.00248	+0.77	-0.07	1.6
4b	37	23	21	659	1542	2674	0.00224	+0.61	-0.35	1.0
4c	33	15	15	729	1626	2760	0.00216	+0.39	-2.15	1.1
4d	29	11	13	903	1856	3076	0.00207	+0.76	-2.60	0.4

[1] Rotovisco Viscometer, PK II spindle, scale reading at 54 speed
[2] 32 Hours
[3] 2 watts - 24 hours

This section has shown that the vehicle components can indeed influence resistor properties. Other experiments showed similar trends, but were not included for simplicity. Because the mechanisms of oxidation and reduction are so complex, it should not be assumed that these same trends will apply to other resistor formulations. In fact, experience indicates that different trends will probably be observed. Therefore to effectively utilize vehicles, experimentation should be performed with the specific resistor formulation to observe the trends there.

The next section will show how the process by which the powders are dispersed in the vehicle can also effect the final resistor properties.

The Milling Process. The pigment of viscous printing inks is usually dispersed in the vehicle by a roll milling process.[28] The paste is sheared between three to seven rotating parallel cylinders, and the shearing force transmitted through the vehicle tears agglomerated chunks of particles apart and disperses them throughout the vehicle. Roll milling should not normally be considered an attrition process; the clearance between the rolls is normally about 10 times as large as the ultimate particle size. In general, the more viscous the vehicle, the greater the shearing force on the aggregated chunks, and therefore the more efficient the milling.

This milling process changes resistance values, T.C.R.'s, frequency dependence, and may improve load characteristics. T.C.R.'s of silver-palladium oxide-glass resistors are commonly driven more negative, and thus formulations can be made with T.C.R.'s very close to zero. Resistance of low R formulations[32] (in the order of $10-150 \Omega/\square$, as the compositions in Table 4) increased with milling by as much as an order of magnitude for a drastically milled paste as compared to an unmilled one. Milling tends to decrease the frequency dependence, and has halved the change in resistance under load in such formulations.

Milling obviously provides a very practical means for improving the control of various lots of resistor paste.[55] Even if the raw materials have varied from lot to lot, each lot can be made identical to the others by providing an appropriate number of compensating milling passes. For non-critical circuit applications, this could permit utilization of untrimmed resistors, with reproducibility in the $\pm 5\%$ range. To do this effectively, the mill must be calibrated in some fashion so that it can apply similar shearing forces on each paste (assuming of course that the viscosity from lot to lot remains relatively constant). It is the author's opinion that the calibration of mills by using feeler gauges between the rolls is not adequate; use of either hydraulic or pneumatic pressure controls (which push the rolls together at a predetermined pressure), coupled with accurate temperature control of the rolls, would be far superior. For each type of formulation, the optimum number of milling passes and milling pressure would probably have to be determined independently. For example, if T.C.R. is a particularly critical factor, the number of proper milling passes for zero T.C.R. in the desired temperature range should be determined experimentally, and then the formulation slightly modified to bring the actual resistance value into the desired range for that number of milling passes. In similar fashion, if T.C.R. is not a particularly critical aspect, the number of milling passes can be varied to bring the resistance to the exact value desired.

The low resistance (ten to several hundred ohms per square) silver-bearing formulae which utilize palladium oxide doped with lithium show much greater changes with milling than do formulae in the middle resistance range (several thousand ohms per square). The more silver present in the formula, the greater the susceptibility to such changes. Typically, the low resistance materials increase resistance with more milling, while very high resistance formulae have been observed to decrease in value with better dispersion.

Another method for controlling the reproducibility of resistor paste involves mixing various lots which differ in resistance value. Many laboratory evaluations on small batches indicate that lots of the same formulation can be mixed to achieve an intermediate value with a very high degree of precision. Thus, it is often wise to avoid making up a single large amount of resistor paste, but instead to make up several smaller lots, which can be intermixed to an exact desired value. In fact, having on hand a single lot of low value and a single lot of high value to be added as necessary to modify other lots, would be a distinct production advantage and eliminate a large amount of batch-to-batch fluctuations. This can be done by the ink maker, or by the user if the latter has adequate mixing facilities.

Control of the Milling Process. It is helpful to control the milling process by some form of quick test on the material being dispersed, so that the appropriate number of passes can be determined. Three such methods were investigated: grind gauge reading, optical reflectance of the paste, and a quick-fire resistance value. The grind gauge reading did not turn out to be a practical predictive means; the relationship between the resistance and grind gauge reading turned out to be erratic (Fig. 4). At the smaller particle sizes, which are desired for proper rheological characteristics and good screening, the resistance value changes too rapidly to be very useful. The grind gauge value is at best an approximation -- rarely accurate to better than ± 0.15 mil of particle size. An attempt was made to control the resistance values of three batches of paste by milling to a desired grind gauge value, but the wide spread of resulting resistances indicated that this was not a good method.

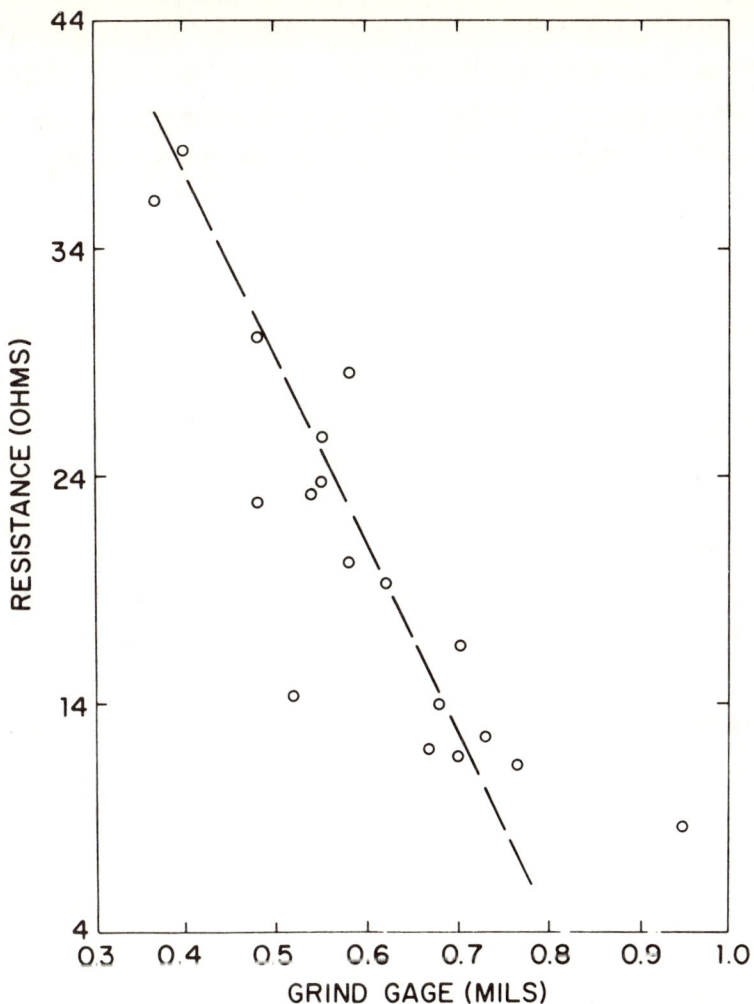

Fig. 4. Correlation between grind gauge reading and resistance.

As particles become smaller, coatings pigmented with them generally become darker, particularly with light-absorbing dark powders. The optical reflectance of printed resistors might therefore be related to resistance values, but this was also found (both with wet and fired resistors) to give rather erratic relationships. A linear relationship between dried resistor prints and final resistor values was found, but the spread of data was sufficiently wide to preclude the use of this test as a predictive tool.

The remaining calibration test which was examined was an actual firing cycle. However, actual resistor firings would take too long to be of practical use for calibration purposes (the inkmaker with a batch of paste on the mill does not want to wait an hour or more before a resistor can be screened, fired and tested to determine how many more passes that paste should receive). Therefore, a ten minute quick-fire at 750°C was tried; where the sample was thrust into a furnace at that temperature. As shown in Fig. 5, this does indeed permit (in a realistic time scale) a good indication as to how many more milling passes are needed. With such a milling method -- whereby a standard number of passes is taken on the paste lot, a quick firing performed, and a judgement made as to how many more passes are required to bring the lot exactly into value -- minor discrepancies in the raw powders used for preparing the paste may be overcome and hidden. To prove this, many examples of different raw materials were brought to within 10% of nominally chosen resistance value. This was true even when extremely wide variations in the raw material preparation were deliberately made, just to see how much latitude in raw material preparation and mixing could be hidden by control of the milling process. Figures 6 and 7 show typical examples of the effects of milling upon silver-palladium oxide-glass resistors, as well as indium oxide resistors.[33] One can see that a calibration chart can be prepared so that if a batch is received which produces resistor values out of specification in a manner that can be improved by milling, it can be readily brought into specification.

It should be realized that investigations of this type are sometimes rendered uncertain by the effects of screens, pressures, operators, firing cycles, and electrodes.[45] The errors or variations imparted by these factors can cloud results. Similarly since rheology of pastes effects resistance by controlling paste transfer, it is often difficult to isolate intrinsic resistance properties from those differences created by transfer characteristics. For example, increased milling tends to decrease paste flow as the agglomerates are broken down and particle surface area increases. Some of the resistance increase of such milled inks might therefore be due to dispersement of pigment as well as the resulting transfer change. It would indeed be desirable to develop a resistance test which is not dependent upon screening (i.e., testing a controlled volume of paste in a cavity), but the author is not aware of any successful tests of this type. Another possible interesting test method which was not performed, but which might be of interest to experimenters in this field, would be the freezing of paste, and then the cross-sectioning of the frozen sample so that the dispersion could be studied and isolated from the transfer characteristics.

Conclusions. The more glaze resistors are studied, the more it becomes apparent that there are very few aspects of formulation or processing which do not potentially have an effect on the final resistor properties. Both the paste supplier and the paste user should be aware of these possibilities, and thus take the proper steps to ward off problems before they occur. The normal tendency is to control every step of the process as tightly as possible. However, a logical alternative is to choose some of the processes which can be deliberately varied-- to compensate for uncontrollable variations in other parts of the process. Probably the most viable steps of the process to vary are the ink milling and the resistor firing.

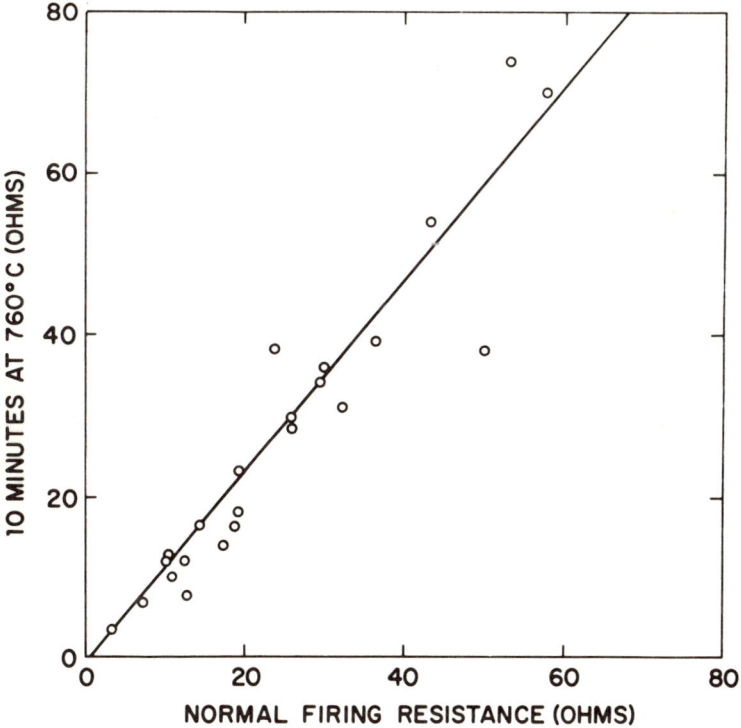

Fig. 5. Correlation between "quick - fire" resistance and normal processing resistance.

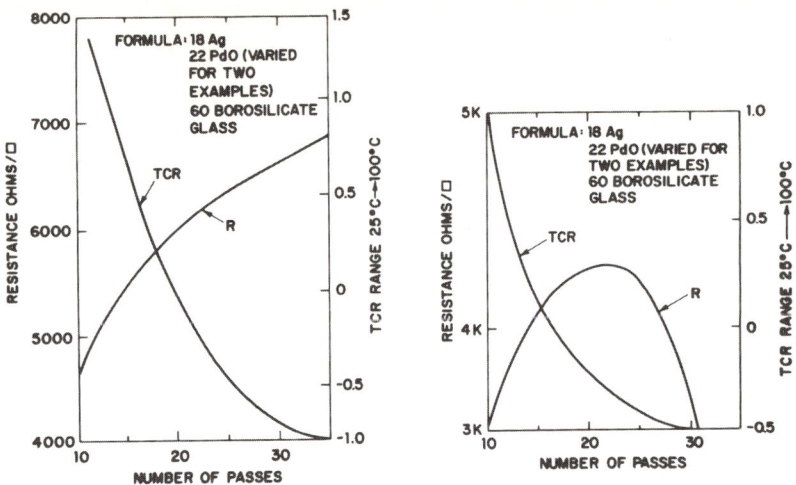

Fig. 6. Effect of milling on two different medium resistance Ag:PdO glass pastes.

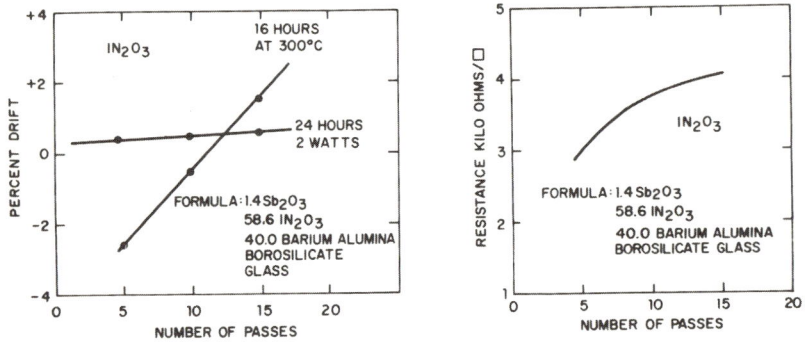

Fig. 7. Effect of milling on various electrical properties of In_2O_3 paste.

Chapter 6

CONTROLLED COLLAPSE REFLOW CHIP JOINING

Introduction. The component industry has developed a number of processes for joining semiconductor devices face down to carrier modules,[56-58] but these can generally be classified within two principal categories -- dry and wet process. The dry processes can be represented by essentially solid state metallurgical interactions, such as thermocompression or ultrasonic bonding.[59,60] Wet processes generally take the form of solder reflow, such as in IBM's SLT module fabrication.[61] This chapter is concerned with a particular wet solder reflow process, called "controlled collapse" and employing ductile metallic pads which are heat-reflowed to form the connecting joints.[62] Significant economic advantages are attainable, as has been demonstrated through recent use of the process with the same automated equipment used for processing SLT modules.

Face-down ("flipped") semiconductor devices have often used rigid contacts (such as copper balls) to ensure support during joining, because diced devices might be electrically shorted if their contacts were to collapse when heated and allow the device edges to contact any conductive material.[63] However, if collapse-induced shorting is prevented, soft solder pads provide significant advantages since their ductility permits them to tolerate considerable stress during thermal cycling and leads to reduced strain and fewer contact failures. Furthermore, since solders are both malleable and capable of distortion (particularly when molten -- as during chip joining), some lack of planarity in the pads or lands to which they are joined can be tolerated in the chip joining process. Thus multi-pad devices of significant size and complexity can be handled when the problem of edge shorting is eliminated.

The first section describes and compares several variations of the controlled collapse process; the second section provides a summary discussion of the module materials involved and shows how the screened conductors can be modified for special uses.

Process. Before discussing the process, it might be helpful to describe the components being considered. The silicon devices, or chips (Fig. 1), have electrical contact pad areas which are metallized with solderable materials (e.g., a layer of chromium to bond to the glass, a layer of copper for solderability, and a very thin layer of gold to protect the copper from oxidizing.[63] The solder itself is then deposited on these metallized pads.[64] Lead-tin alloys with high lead concentrations are used primarily, but other alloys are feasible and, in certain applications, may be necessary for metallurgical compatibility. The solder can

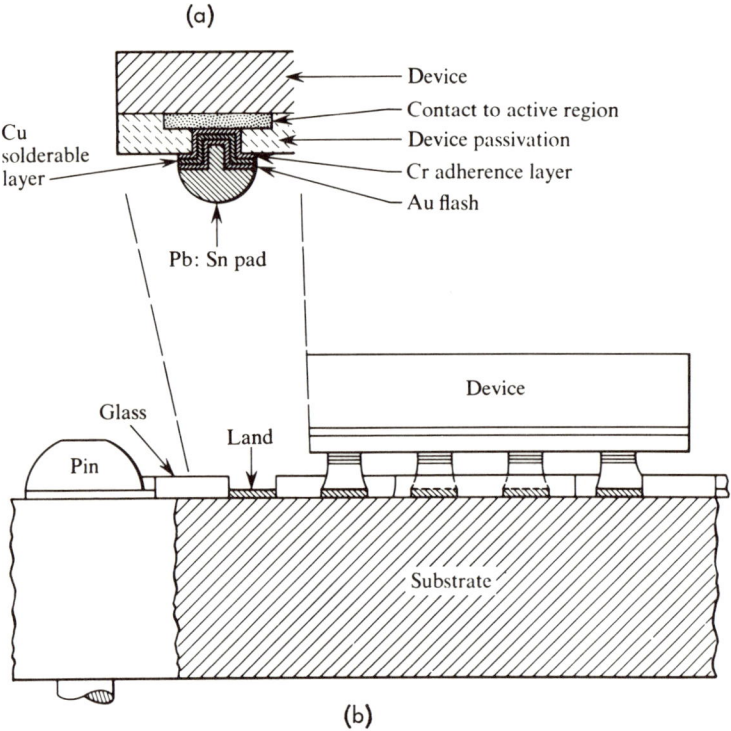

Fig. 1. (a) Side view of a device; (b) side view of a device on a substrate (dam method).

be applied to the metallized pads by vacuum metallizing the lead-tin alloy in an oversized area around each pad and then heating it to form a hemispherical ball by reflow. Since the solder does not wet the oxide surfaces around the periphery of each pad during this first reflow operation, relatively large hemispheres can be formed. Typically, they are in the order of 4 or 5 mils in diameter (Fig. 2). The devices and substrates are made separately and jointed to each other in reflow furnaces containing either inert or reducing atmospheres, such as nitrogen, argon, or forming gas.

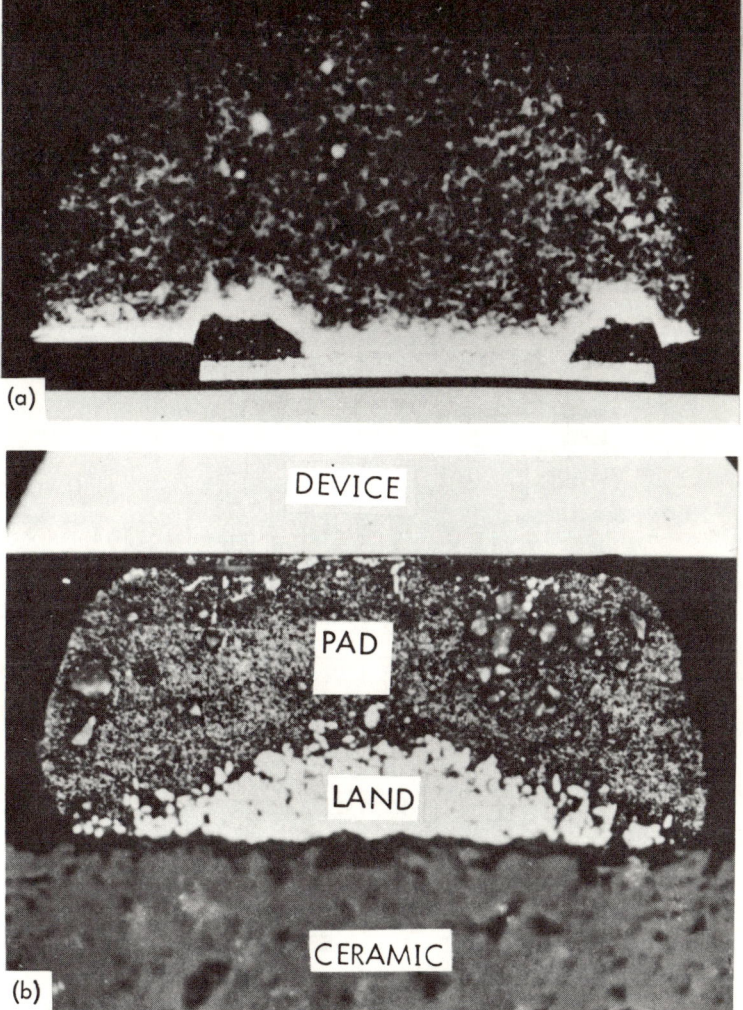

Fig. 2. (a) Cross-section of a "wetback" solder pad; (b) device joined to ceramic with solder pad. (Photographs by M. Ricker.)

The process involves restricting solderability to those spots on the substrate at which the chip pads will make contact, so that solder cannot run along the lands and cause the chip to collapse and thereby short in the land solder on the substrate. Many ways of accomplishing this are possible, but we will restrict most of the discussion to two typical process types, called "dots" and "dams" and to some variations of these (see Table 1 and Fig. 3).

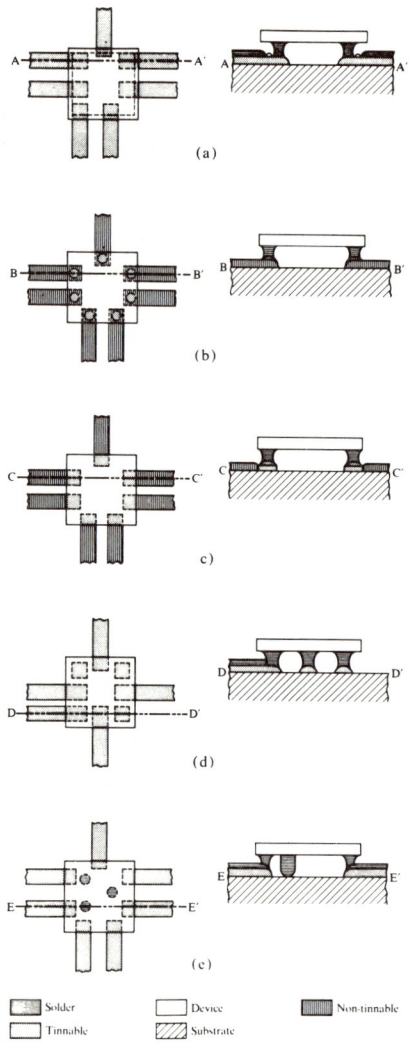

Fig. 3. Illustration of the methods compared in Table 1. (a) Dams; (b) dots; (c) overlap; (d) extra pads on module; (e) extra pads on device.

<u>Dots</u>. With this method, a circuit pattern is screened on the ceramic substrate with a non-tinnable electrode paste (discussed later) which is dried and fired. The tips of the fingers to which the device pads will be joined, as well as any areas which must also be tinned to ensure a reliable interconnection (such as the area around pins), are then screened with normal, tinnable electrode paste.

TABLE 1. Comparison of methods.

Method	Procedure	Advantages	Disadvantages
Dams (a)	1. Screen, dry, fire tinnable conductive lands. 2. Screen, dry, fire or cure solder stop-off dams across finger ends. 3. Pin and tin module. 4. Join devices. (Example: glass dams on Ag:Pd lands)	1. No shorting of adjacent lands. 2. Same dam material can be used for crossovers (if glass), or overcoating (if polymer). 3. No new metallurgical compatibility required.	1. Tinning tends to give some difficulty. 2. Resistance may be higher under dam.
Dots (b)	1. Screen, dry, fire non-tinnable lands. 2. Screen tinnable dots and other tinnable areas (as pin areas), dry and fire. 3. Pin and tin modules. 4. Join devices. (Example: gold non-tinning lands, Ag bearing dot.)	1. Potentially better tinning 2. Slightly easier registration. 3. Minimal silver migration. 4. Potentially better conductance.	1. Adhesion of dot. 2. Possibility of dot bleeding out to cause electrical shorting. 3. Proper metallurgical systems are difficult to optimize. 4. If crossovers are required, this process does not help.
Overlap (c)	1. Screen, dry, fire isolated tinnable regions. 2. Connect these with non-tinnable lines, fire. 3. Pin and tin modules 4. Join devices (Example: Pt or Ag:Pd tinnable regions connected by Au, Ag:Pd; or Au:Pt non-tinning lines.)	1. Less adhesion compatibility required than with dots. 2. Simpler to use inert finger ends (as Pt).	1. Difficult to screen well. 2. Registration is quite critical.
Extra pads on module (d)	1. Screen, dry, fire all tinnable regions. 2. Pin and tin modules. 3. Join devices. (Example: any tinnable land.)	1. Only one screening 2. Better tinning yield.	1. Sacrifice of silicon real estate. 2. Requires balanced format. 3. Possible capacitive effects.
Extra pads on device (e)	1. During device pad operation, deposit extra pads. 2. Prepare module in normal way. 3. Reflow devices. (Example: Pb:Sn pads)	1. Very low cost — no additional steps.	1. Requires space for extra pads on insulated portion of device. 2. Requires balanced format. 3. Possible capacitive effects. 4. Does not provide the self-alignment of other methods.

The module is again fired to sinter the tinnable area and bond it to the underlying non-tinned land. Any screened passive components that are required for module fabrication are then added and fired. Normal pinning and dip tinning operations follow. During the dip tinning operation (which is generally done in 90% Pb, 10% Sn solder), the solder adheres only to the pin areas and to each finger tip, but not to the non-tinnable land sections (Fig. 4). Devices with solder pads (and any other necessary components such as chip capacitors) are then reflow-bonded to the isolated tinned spots. Since the chip edges extend beyond the solder and since surface tension supports the device, collapse does not create shorting at the device edges.

Dams. With this second method, the ends of normal screened and fired tinnable lands are delineated with a non-tinnable barrier. After firing or curing the barrier, or dam (which can be done simultaneously with the firing of other passive components, if desired), and after pinning, solder is deposited on non-covered land sections. As in the dot method, this allows isolated solder spots on the finger ends to receive the device solder pads (Fig. 5). The untinned area again prevents collapse-induced shorting since the chips remain suspended several mils above the lands; the untinned areas do not touch them. Many types of solder stopoffs can be used for this process and some of these will be discussed. This method has received particular emphasis at IBM.

Overlap. This method may be considered the inverse of the dot process; instead of depositing the untinned lands first, the tinnable sections are deposited and fired and then interconnected with non-tinnable land patterns. A significant difficulty here is the maintenance of good land definition at the overlap region. Electrode paste can be squeezed under the screen or mask (which might be lifted by the previously printed and fired tinnable areas), and the dimensions can be difficult to control. Thus, although this method is certainly feasible, the others are considered somewhat easier to employ.

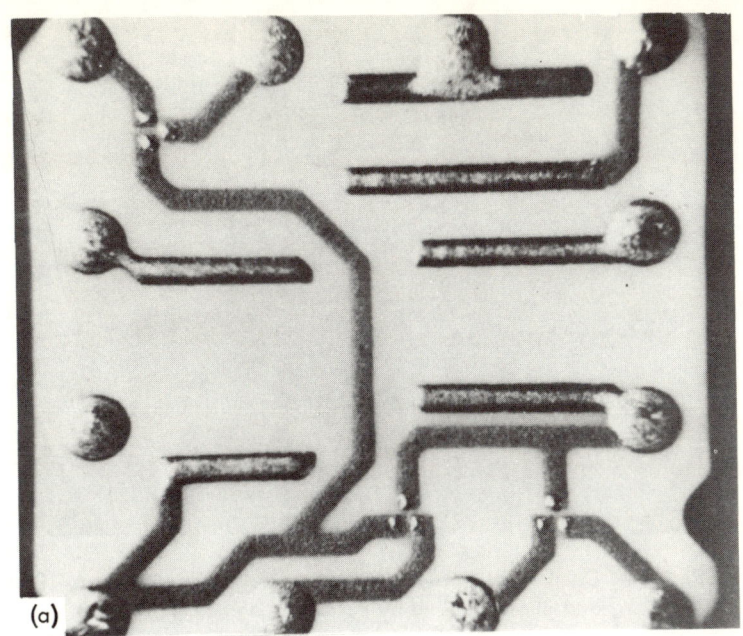

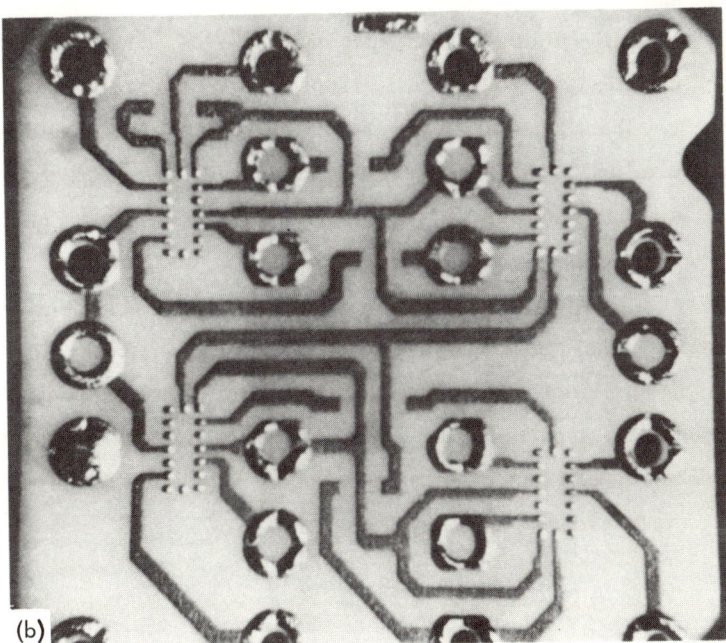

Fig. 4. Two modules with combinations of tinnable and non - tinnable lands, and "dot" configurations at ends of lands.
(Photograph, Fig. 4b, by L. Otten.)

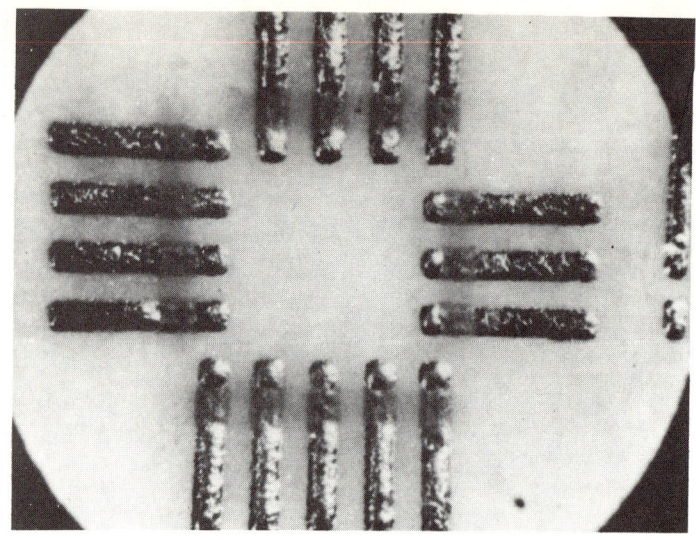

Fig. 5. Dams (cured polyimide) on Ag:Pd lands — tinned. (Prepared and photographed by F. Glenn and K. Puttlitz.)

<u>Isolated Lands</u>. Not all of the lands must have this solder-wettable area defined; if just a few are so defined, perhaps three or four sites out of twenty or so on a device, the surface tension of those isolated pad areas is sufficient to maintain separation. The isolated areas on the modules do not have to provide electrical contact, but can serve merely as mechanical supports (Fig. 6). It has been calculated and demonstrated that the surface tension forces in the scheme described here can support considerably more weight -- at least 30 times more-- than they do.[65] There are possible disadvantages to this variation, however, including some sacrifice of electrically useful silicon area if more pads are used than are necessary to establish the electrical contact, possible capacitive effects from these extra pads, and possible lower chip elevation than would exist if all pads were to serve as supports.

<u>Extra Pads</u>. Extra pads can also be placed on the chip, with no corresponding wettable land portion on the module (Fig. 7). Since the solder on the pads cannot wet the ceramic module surface, the surface tension of these extra pads will maintain the chip at an adequate distance from the substrate.[66] Again, however, capacitive effects must be considered.

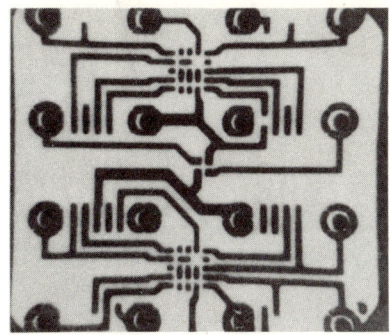

(a)

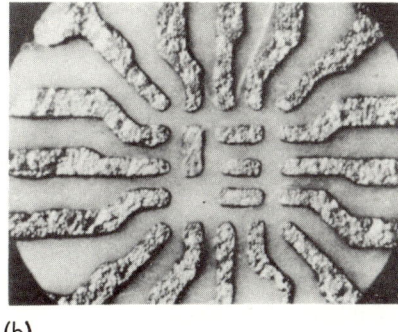

(b)

(c)

Fig. 6. (a) Land pattern (dried) using extra pads which can prevent collapse;
(b) extra pad configuration, etched from copper (screened resist);
(c) isolated pads on ceramic.

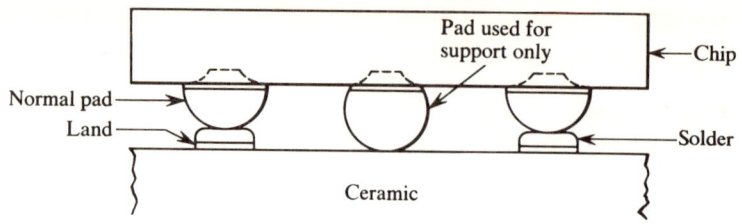

Fig. 7. Side view of extra standoff pad on device, used to prevent collapse.

Other techniques for using extra, non-electrical supports -- either on the module or on the chip -- require careful choice as to number and location, or tipping will occur. When supports are used, the potential problems in pad planarity become more important than with the other systems discussed previously.

<u>Common Characteristics</u>. These processes have the following similar characteristics:
1. The first three involve two screening operations -- one with a tinnable conductor and the other with a non-tinnable material. All are compatible with normal semiconductor, thick film module processing, with only some minor modifications required for format changes and slight variations in process detail.
2. Registration and alignment of the tinnable and non-tinnable sections is relatively critical in order to achieve the controlled collapse of the device. When tinnable spots of about 4-8 mils are used with device pads of about 4-5 mils in diameter, the chips are normally suspended between 2 and 4 mils above the lands, providing good stress relief through the long ductile members.
3. The devices are, to a large extent, self-aligning on the limited solder areas, so that when contact is made between a part of each solder pad on the device and the land solder, the chip will float into position during the reflow process. Considerable misalignment in chip placement registration can be tolerated in some geometries due to this correcting mechanism.

4. Irregularly shaped pads, and even crushed pads, on the chips can be used, since during the reflow process these pads re-form to hemispherical shape and can be joined as well as initially uniform pads. Thus defects due to prior electrical tests are eliminated, and the probes do not damage the pad joints, as is possible with rigid pads.

5. Chip joining yields above 97% have been routinely demonstrated; chips can be replaced as required. Joint strengths are very good, typically 50 gm per pad in tensile strength, equaling reflowed joints made in other ways with hard device pads. High reliability has been demonstrated, even with large devices.[67]

6. The solders found to be particularly successful have been 95% Pb, 5% Sn for the device pads, joined to 90-10 solder on the lands. The devices are held down with soldering flux during the reflow, which can be done at about 340°C for this metallurgical system. More rapid reflow at temperatures close to the solder liquidus will minimize metallurgical problems.

7. The methods are applicable to multiple devices on a single surface and to devices with many pads. Using screening technologies, module lines and spaces of 4 mils can be manufactured; some modifications in technology might be required to reduce line dimensions and separation beyond that for production purposes.

Although this particular disucssion is directed toward thick film technology, these concepts can also be extended to thin film technology and even to normal copper-clad printed circuits, if the thermal expansion differences of the materials are not limiting (See Fig. 6b as an illustration of etched lands).

<u>Discussion of Process</u>. These methods have been shown to permit wide flexibility in module construction. Devices with pads approximately 5 mils in diameter have been joined equally well in configurations ranging from 3-pad SLT fingers (which are approximately 15 mils wide) to devices with many pads on centers as close as 8 mils. For manufacturers who produce modules using SLT-like processing, the impact on production equipment would be minimal, since the compatibility is obvious.[68]

The joining of any two bodies with dissimilar thermal expansion characteristics can potentially cause a reliability problem.[69] As devices become larger and larger, the difference in thermal expansion between the device silicon and the alumina of the ceramic tends to become increasingly important. However, due in part to the ductility of the solder pad joint, devices of the order of 100 mils square have been shown to be highly reliable after many hundreds of cycles through wide temperature excursions, such as from -40° to $+150^\circ$C,[67] the upper limit of size has not been established. Theoretically, the farther the device is suspended from the ceramic -- that is, the longer the ductile beam -- the smaller the stress that will be transmitted during such thermal stress periods to the critical joints where the pads meet the device. These elongations can be achieved by the proper choice of volume of pad material and size of the tinnable areas, permitting the use of large devices.

The tensile strength exhibited routinely by these joints has been of the order of 30 to more than 50 grams per joint; such strength is well beyond that required by normal device joints. Typically, for example, thermocompression wire bonds average only approximately 2 to 7 grams per one-mil wire when the wires are pulled off.[69]

The reliability of "flip chip" devices has been shown to be excellent, as represented by billions of hours of field usage, coupled with correlation to laboratory stress tests.[70] SLT devices which use rigid copper balls as pads were reported to show a ball failure rate of about 0.00004% per thousand hours at a 90% confidence level. This enormous amount of field data is not yet available for the all-solder joints discussed here, but laboratory stress tests indicate that a considerable improvement in joint reliability can be expected from the ductile joints.[67]

Pilot quantities of tens of thousands of modules using devices with more than 10 pads each have shown chip joining yields above 97% with the dam method. One hundred percent visual inspection of 500,000-600,000 joints showed no cold joints. These yields are equal or superior to any known joining scheme that has been developed to the point of pilot production. The very nature of the controlled collapse mechanism permits somewhat more flexibility in temperature, time, placement, and even dimensions than do other chip joining methods. This wider latitude in processing is one of the reasons yields are very high.

If modules containing active devices are cooled by direct immersion in liquid coolant,[71] better thermal dissipation can be achieved with flip-chip devices such as these than with back-mounted devices; the liquid can circulate around all of the sides of a flip-chip device. The heat dissipation is particularly effective if thin metal heat sinks are fastened to the backs of the devices, expanding the area of the thermal path. Thus, the advantages of economy, reliability, ease of replacement, and yield for these processes can be extended to high power devices.

Another consideration is the replacement of devices on multi-chip modules. This has often been demonstrated, and devices joined even after the third or fourth replacement have shown strength equivalent to that of the initial devices. These "controlled collapse" chips do not necessarily require localized heating to replace adjacent chips -- the entire module can be reheated.

Materials. The materials described in this section have been used to demonstrate controlled collapse processes using existing SLT module fabricating equipment,[68] which, within some limitation of line widths, conductivity, and circuit card compatibilities, demonstrates unique advantages of economy, flexibility, and reliability. This section will discuss the general types of screening materials and some considerations relating to their formulation, limitations, and performance. The following types of materials will be discussed: (1) tinnable lands; (2) non-tinnable lands; (3) solder stop-offs (dams); and (4) dots. The experimental data and conclusions described here were obtained under the same conditions as in previous chapters.

Tinnable Lands. Tinnable lands are used directly as joining sites and to create tinnable dots. Since the latter may create special problems with respect to compatibility, they will be discussed later, Tinnable lands are used routinely by circuit manufacturers, but it was expedient to formulate special products to fill particular needs. Similar characteristics are required from these materials

113

for ductile pad processing as would be desirable in normal thick film processing: good adhesion, high conductivity, excellent screenability, low cost, high tinning yield, etc.

Conductivity. In the dam method, the solder layer is interrupted on the lands by the dams, and a high resistance section can result in the untinned land resistance is high. For circuits which do not permit such areas of high resistance, the electrode formulae discussed in chapter 2 which contain a metals ratio of approximately 80 parts of silver to 20 parts of palladium is useful. Although there is extensive literature on the migration of pure silver lands, the alloying effect of the palladium drastically reduces such migration, and permits these electrodes to be used in module fabrication. As an example of the importance of the land resistance, even under a small dam section (depending on the line width), such a silver-palladium land will either show no resistance change at all under a dam approximately 5 mils wide, or an increase in resistance on a 5-mil wide land up to about 0.05 ohms. In this format, commercially available gold-platinum lands would add about 0.2 to 0.7 ohms for the untinned section.

Depending on screening conditions, all pastes can produce a relatively wide range of land resistances. Thus, if wider or thicker lands are used, gross conductance increases can be attained, as shown in Table II. By intelligent use of screens or masks, such as by increasing emulsion or mask thickness, eliminating the mesh or maximizing the size of the mesh holes, and adjusting squeegee angles and pressure etc., improvements can be produced.

Table II. Resistance and Land Height Measurement for Different Screening Conditions.

Metal	Screen	Firing	Untinned Resistance, μ/inch			Line Height, microns			line width
			3-Mil	4-Mil	5-Mil	3-Mil	4-Mil	5-Mil	
Ag:Pd	325 mesh	750°C	29.5-38	13.5-16	6.4-8.5	6-8	7-10	10-13	
Ag:Pd	3-mil open mask	750°C	4.6-7.0	2.0-4.0	1.3-1.5	12-14	21-23	24-26	
Ag:Pd	4-mil open mask	750°C	6.5-7.0	1.8-2.2	1.1-1.2				
Cu	Mesh mask	air 750°C, then forming gas 1000°C	3.2-6.5	2.8-6.5	1.0-4.0				
Cu	3-mil open mask		0.50-0.60	0.30-0.45	0.28-0.35	7-12	15-17	18-20	
Cu	5-mil open mask				0.34-0.5				
Au	3-mil open mask	750°C	3.0-5.5	2.2-2.5	1.9-2.1	17-23	22-28	25-30	
Au	4-mil open mask	750°C	5.0-6.0	1.6-2.3	1.3-1.6	10.5-13	17-28	25-27	
			5-Mil	10-Mil	15-Mil	5-Mil	10-Mil	15-Mil	line width
Ag:Pd	200 mesh	750°C	5.3	1.4	0.8				
Ag:Pd	2-mil cavity mask	750°C	5.3	1.4	0.9	10-12	24-25	23-24	
Ag:Pd	3-mil cavity mask	750°C	4.0	1.0	0.6	13	21-32	23-32	
Au	200 mesh	750°C	1.6	0.4	0.23				
Au	2-mil cavity mask	750°C	1.5	0.43	0.24	9-11	14-23	24-29	
Au	3-mil cavity mask	750°C	1.3	0.38	0.22	9-12	16-18	20-26	

Land resistance is partially controlled by the interaction of the solder and the land, and also by the amount of solder deposited by the tinning process. The solder height can be increased by rapid withdrawal rates from the molten solder, lower solder temperatures, and the presence of intermetallics which raise the solder viscosity. Of course, in any tinning operation, some compromise may be required by balancing such defects as bridging, unwet areas, and intermetallic brittleness with the desired high solder height. Since most tinnable materials -- and particularly silver or gold bearing materials -- dissolve somewhat in molten solder, the amount of land material present is extremely important. Thin layers obviously lose a larger proportion of metals into the solder than thicker layers. Thus, for example, 5-mil wide silver-palladium lands typically show a slight resistance increase after soldering, while 10-mil lines and over generally show decreases in resistance after tinning when the lines are screened through at least 200 mesh. With such screens, narrow lines are usually shallower than wide lines, as observed in the second chapter. However, such relationships are less discernable when the lands are screened through metal masks; thick 6-mil lines can actually improve conductance after tinning. Intermetallic formation is also pertinent.

<u>Adhesion</u>. Adhesion of the lands is normally important in two contexts: The reliability of the contacts, and the ability to replace devices without pulling the lands from the modules. Experience indicates that for land fingers, any electrode material whose adhesion equals or exceeds 2500 psi would be adequate if chip replacement is performed by re-melting the pads. However, this adhesion may not be adequate if severe stresses are applied to external contacts such as pins or leads. The module manufacturer should not depend on the electrode-to-module adhesion for reliability of his external module interconnections when relatively low temperature ($750^\circ - 1000^\circ$C), air-fired formulations are used. Of course, active metals such as moly-manganese, tungsten, etc. typically produce considerably higher adhesion, and thus tend to be more useful under severely stressed conditions. These require firing in reducing atmospheres at very high temperatures, and must be plated to be made solderable, thus increasing the complexity, cost, and controls required for production.

Adhesion is a complex interrelationship of the ceramic surface, the metals in the electrode formulation, and the glass frit (if included). In general, increasing the concentration of frit tends to increase the adhesion, although in certain examples this relationship was shown not to hold. In some formulations, a maximum adhesion occurs at a particular frit concentration. In general, the more frit added to the formulation, the less tinnable the formulation tends to be. The denseness of the sintered metal is also an important factor; in fact, it may be one of the critical aspects of what is commonly thought of as adhesion.

It can be hypothesized that when the land is fired onto the ceramic surface, the frit in the land acts as a flux and permeates the glass binder which is holding the grains of the ceramic body together. Not only does this bond the frit to the ceramic, but electron microprobe analysis indicates that some of the glass from between the ceramic grains may then ooze up and cover the surface of the ceramic Thus, essentially a new interface can be formed which previously did not exist. For the frit to enter the ceramic structure, a region immediately above the base of the electrode should be slightly depleted of the flux that was originally there. This should be the weakest section of the electrode, and on certain types of formulations -- such as the silver-palladium electrode referred to before -- this thin layer, containing some small amounts of entrapped metal particles, is predominantly the point at which the electrodes yield.

If such an analysis truly represents what occurs, the failure might well be termed a cohesive failure rather than an adhesive one. Of course, there are some cases when the cohesion of the system can be increased and improved to the point where failure occurs in the solder layer, and in such cases, this postulated mechanism would not hold. For example, if the ceramic surface is glazed, instead of depleting the flux from the electrode, the glaze actually enters the electrode from beneath and supports its cohesive structure, and therefore could increase the strength considerably. This is true, of course, only if the cohesion of the electrode above the permeated glazed region is sufficient to supply adequate strength to the structure. For example, experiments show that the silver-palladium electrode has sufficient cohesion on glazed surfaces to permit very good strength, while commercially available gold-platinum yields cohesively only a few hundred psi higher on a glazed surface than on a non-glazed surface. Of course, by varying the materials and processing conditions, this description may not hold universally.

All of the electrodes described here showed at least 2500 psi in tensile strength. The strengths of chip joints were examined by pulling the chips upward tension at 0.02"/min. Typically, 4-mil diameter chip pads on 6-mil tinnable land areas produce in the order of 50 grams per joint or better when the appropriately matched metallurgical systems are used. This strength was shown to be more than adequate with ductile solder pads. However, brittle intermetallics formed by interactions of the solders and other metals in any system can produce a more brittle joint than might be desired. Therefore, barrier metals, such as nickel, can be used to avoid such intermetallic formation, and the processing temperature and time can be adjusted to minimize this effect. Optimization of the metallurgies and the processing can indeed produce reliable ductile joints in these processes.

Non-Tinnable Lands.[72] Non-tinnable lands can be used to connect tinnable areas and to produce conductive solder stop-offs (see Fig. 4). The pastes are formulated similarly to tinnable lands -- containing mixtures of powdered metals, frit, and organic vehicles to make them screenable -- except that small quantities of finely dispersed non-tinnable particles are included to prevent interaction of the lands with solder. Lands of this type do not accept solder or erode in solder, and resistance changes after tinning are negligible. Since these lands are not tinnable, metals alloys and frits which could normally not be used for tinnable lands become practical. For instance, nontinning gold lands with good adhesion were formulated, although tinnable gold lands are not very successful since gold dissolves readily in lead-tin solders. Thus, non-tinnable lands can potentially be made even more conductive and adherent, and even -- in the case of silver -- less migratable then tinnable lands. For instances, it is known that magnesium is useful as an inhibitor for silver migration of screened lands.[73] Magnesium normally might interfere with the tinnability of such lands, but if the lands are designed specifically to be non-tinnable, this is not a detriment. Of course, such a system might require firing in an inert or reducing atmosphere to avoid oxidizing the migration inhibitor, which increases cost and conplexity.

Another method of making normally tinnable lands non-tinnable involves firing the normally tinnable land on a ceramic substrate with which it can interact during firing. When the glazed material from the ceramic surface enters the structure of the electrode, the lands are made non-tinnable. Such ceramic surfaces generally contain relatively low softening point glasses which perform such interaction.

The non-tinning agents are used in relatively small amounts, typically 1 to 4% by weight of the formulation. The actual amount depends primarily on three things: the type of non-tinning agent, the kind of frit in the paste, and the metals in the paste.

Type of Agent. To make the non-tinning filler most effective, the finest dispersion possible is required. Laboratory studies show that the most predictive measurement is the oil absorption value of the powder (Table III). Less correlation of the amount of material needed to prevent tinning was obtained with particle size or surface area measurement. High oil absorption value (or small particle size) is desirable. As representative of the kinds of materials which are particularly successful, oxides with high decomposition temperatures (preferably over 1000°C) and high free energies of formation are generally best. Collodial powders such as the oxides of silicon, aluminum, indium, chromium, etc., were used. For most applications, however, collodial silica (such as cabosil) is particularly effective. In addition, organo-metallics, polymers, or salts which decompose during firing to form such oxides were also used advantageously.

Table III. Non-Tinning Agents.

Example of trends in a group of siliceous materials - 3% in Ag:Pd.

Agent	Composition	Tinning	Surface Area (m^2/g)	Average Particle Size (microns)	Oil Absorption* (lbs oil/100 lbs)
syloid AL-1	colloidal silica	tins	700	10	90
syloid 75	colloidal silica	sporadic	340	2.9	225
syloid 162	colloidal silica	sporadic	260	10	145
syloid 224	colloidal silica	almost none	300	3.3	300
cabosil	colloidal silica	none	175-200	0.02	250-300

*Manufacturer's Data.

Electron microprobe analysis of non-tinning silver-palladium lands showed that the collodial particles were well dispersed throughout the fired land, and that they frequently interrupted the surface metallurgy. This apparently prevents the solder from wetting and spreading on the land surface. The surface tension of the solder probably prevents intimate contact of the solder with metallic regions between the silica islands.

Interestingly, tinning can be restored to the lands by polishing (which apparently smears a Beilby layer of metal over the non-tinning spots). Thus, some care should be taken to prevent abrasion of the fired lands before the tinning operation. It is also enticing to postulate usage of this phenomenon to locally polished sections, where tinning is desired, and thus avoid an extra screening step. Such a procedure was demonstrated successfully, but the tooling might be expensive compared to that required by some of the other methods.

Metals. For inconclusive reasons, different land metallurgies respond differently to this non-tinning treatment. Not only do different metals require different amounts of non-tinning agents, but some non-tinning agents are relatively ineffective with some metals. Table IV shows some examples. For example, cabosil works well for gold, platinum, silver-palladium, and silver-platinum lands, but not for gold-platinum.

Table IV. Examples of Different Non-Tinning Agents with Different Pastes.

			Paste Type	
Material	Au	Pt	80 Ag:20 Pd	80 Au:20Pt
cabosil	OK 1.5% or 3%	OK 1%	needs 3-4%	not 5%
TiO_2	erodes 5%		not 6%	F-G 10%
copper titanate			not 6%	F-G 6%
alumina				
barium acetate	OK 5%			OK 5%

Frits. The frit, flux, or binder in the paste can either enhance or deter the action of the non-tinning agent. If the frit does not chemically interact with the non-tinning agent, it can help stop the tinning of the land, particularly if it is a high-melting, non-reducible glass. However, some low-melting frits hinder the effect of collodial silica, requiring more silica to be used than if no frit were present. This was observed with both silver-palladium and gold non-tinnable paste when Drakenfeld E1313 glass (a lead aluminum borosilicate) was used with cabosil.

Large amounts of appropriately fine particle frits also stopped tinning without any additional agents, such as 10% of Kimble TM7 glass (a relatively non-reducible glass) in either silver-palladium or gold pastes. Another example (combining a frit and decomposable silicone vehicle to prevent tinning) was demonstrated with silver-palladium metallurgy and a combination of barium fluoride and barium borate flux with a vehicle containing Dow Corning R-5071 silicone resin. Chemicals such as barium acetate, which decompose to form high melting oxides, were also shown to be effective.

Characteristics of Non-Tinnable Lands. Any dilution of metals in a paste can be expected to raise electrical resistance unless the paste transfer, metal sintering, or resulting land thicknesses are simultaneously increased. Depending on the volume relationships, the dilution of metals with non-tinning agents generally increases resistance. The densities of the noble metals used in conductors are commonly about 10 to 21, while non-tinning agents are in the range of about 2.5 to 5. Thus, a 1% weight dilution with a non-tinning agent might contribute a 5% to almost 10% volume dilution. Table V compares some of the resistance values for typical tinning and non-tinning formulations. It can be seen that the effects, although real, are not of great magnitude, and are important only in critical formats.

Table V. Comparison of Resistance Values for Workable Tinning and Non-Tinning Lands.

Metallurgy	Type	5 Mils	10 Mils	15 Mils
80 Ag:20 Pd	Tinning	5-8	1.5-1.7	0.9-1.1
85 Ag:15 Pd	N.T.	7.5-10	2.5-2.8	1.5-1.6
Gold	No agent, but erodes	2.0	0.6	0.35
*Gold	N.T.	1.5	0.4	0.25
Gold	N.T.	3.1-4.5	0.6-1.4	0.4-1.1
Ag	Tinning	0.6-0.9	0.17-0.25	0.13-0.17
Ag	N.T.		0.23	0.14

Approximate resistance, 200 mesh screen Ω/inch

*2-mil cavity, generally analogous to 200 mesh

Since adhesion tests on conductors are commonly performed on soldered lands, it is difficult to comparatively evaluate the adhesion of these non-tinning lands. Treatments to make the surface tinnable -- such as polishing, plating, or firing on tinnable material, etc. -- could possibly change the adhesion. For this reason, but with the exception described later, no good adhesion values are available for these formulations. However, probing the lands with scalpel (which can be correlated to strength) indicated toughness and good adhesion for acceptable non-tinning formulations. The collidal particles in nontinning agents tend to increase paste viscosity, increase thixotropy, reduce flow, and increase the tendency to clog the screen during printing. This could be a significant problem since it is coupled with the need for a high solids content to achieve good conductance. Of course, rheological optimizations can be introduced with vehicle changes and proper surfactants to make these formulations usable, but generally they tend to produce pastes which are somewhat more difficult to screen than normal pastes. Fortunately, for many formulations, the amount of non-tinning agent is relatively small, and therefore can be compensated for.

As previously mentioned, regular tinnable lands can be made non-tinnable without special treatment when fired on certain glazed surfaces or on some ceramics which can be fired at low temperatures. The latter are generally the self-glazing types of ceramics. During firing, the glazing from the ceramic surface enters the land structure and makes the surface of the land non tinnable and non-interacting with solder. This is a somewhat more desirable way to achieve non-tinnable lands, since such a procedure can eliminate the rheological difficulties mentioned previously and also produce slightly higher conductivity. However, not all of these ceramic systems are amenable to module fabrications, since they may be more costly, or -- more likely -- because they are inferior with respect to thermal dissipation. Some commonly used electrode systems which must be fired in reducing atmospheres such as tungsten or moly manganese are inherently non-tinnable, but substantially more difficulty is involved connecting tinnable regions to them. Of course, this can always be done with stop-offs and plating, at the sacrifice of added process complexity and cost.

<u>Dams</u>. As previously mentioned, the non-tinning lands can be used as dams or solder interrupts when applied across regular tinnable lands. The main rea-

son for using a conductive dam is to prevent resistance increases in the dam area, but this is only achieved with the proper match of metallurgies in the tinnable and non-tinnable formulae. Since alloying commonly increases resistance, dissimilar metals can be undesirable. For example, if the non-tinning dam contains gold and the underlying lands contain silver, the dammed area may be of considerably higher resistance than if a non-alloying dam were used. On the other hand, if a non-tinning gold dam were placed across a tinnable gold-platinum land, the resistance of the dammed section is almost as good as if the area were soldered. The difficulty in using conductive dams is twofold; the matching of the metallurgies, and the possibility of electrically shorting adjacent conductors when compaction of circuitry requires small spacing between them. Thus, for chips with closely spaced pads, it was found desirable to use glass or other non-conductive dams to delineate the module lands. Although dams made with precious metals may be slightly more expensive than glass dams, the small amount of paste used for these patterns makes this differential rather negligible.

Many types of dams or stop-offs can be used and have been demonstrated -- including glass, metal oxide, polymers, silica, and non-tinning conductors. Several basic attributes are important: the dams should interact as little as possible with the underlying conductive land to avoid conductivity difficulties or tinnability problems at the finger ends; the dams should be of minimal thickness and width to reduce geometric shielding of the finger ends during tinning and to minimize the amount of non-tinned sections; and adhesion and stability should be assured throughout the life of the module. Furthermore, the material should be screenable, not excessively abrasive, reproducible, and it should be able to protect the lands from corrosion (or at least not make them worse). When glass dams are used, they should be fired at or just above the softening point to minimize bleeding into the tinnable land which might interfere with soldering (as occurs with glazed substrates). Complete flowout and glazing of the glass is not necessary -- just enough sintering of the particles to achieve solidity and firmness of the dam to prevent damage during further processing. If the glass pastes contain low solids contents and if fine mesh screens or thin masks are used for printing, the desired sharply defined low profile dams are obtained. Compatibility with a range of processing temperatures is achievable from about 550°C to more than 1000°C.

The number of choices is wide, and the matching of almost any set of module processing conditions is not considered difficult.

Metal oxides, which interact even less with lands than glass, seem to permit better tinning of the finger ends. A particularly effective formulation contains palladium oxide fluxed with 3% of Corning 7720 glass. Dilute dispersions of collodial silicas such as cabosil also produce good solder stop-offs. However, if applied too thickly, cracking results when the modules are thermally cycled. This type of dam would have the potential advantages of low cost, no abrasiveness (as with glass), and no interaction with the tinnability of the underlying lands.

High temperature polyimide resins have interesting advantages in that they can withstand soldering temperatures and thermal cycling, but do not have to be fired. Sharply defined, low profile dams can be fabricated with polyimides, without any deleterious land action. The polyimide area can also be extended, if desired, to provide overcoating protection. Both DuPont's PYRE ML vanish and PI4701 (with 10% cabosil mixed in to provide thixotropy) produce good dams with curing cycles under $300^{\circ}C$.

It would be an intersting extension to utilize glazed resistors -- which might normally be used on electronic modules -- to provide the dammed regions. However, due to the large amount of low softening point glass in normal glazed resistor formulations, several attempts to use such pastes as dams failed due to the uncontrolled bleeding of the glass into the lands at the firing temperatures required to mature the resistors. This bleeding either completely prevented the end of the fingers from tinning, or resulted in tinnable pad sections that were of irreproducible size. However, for non-critical regions, this might be a practical procedure.

Dots. As previously described, dots of tinnable material on top of non-tinning lands can be used to delineate tinnable sections. With this method, compatibility of the two lands and the solder is the most important factor affecting adhesion, contact resistance, and tinnability. In some instances, it is relatively simple to obtain metallurgical compatibility; for example, non-tinning silver-palladium lands with tinnable silver-palladium dots work very well. However, in this system, the non-tinning lands might be more susceptible to silver migration, since they do not have the protective solder layer of normal tinnable silver-palladium lands. Tests show that lead-tin solder on top of silver-palladium electrodes further retards migration.

Better conductance can be achieved with non-tinning gold or copper lands than with silver-palladium. However, since copper requires reducing or inert atmospheres for processing, many technical details must be secured before it could be used (as with the active metals mentioned above), particularly with glass crossovers or resistors which might tend to be reduced. Such processing might also cause extensive equipment outlay for manufacturers not set up to deal with such atmospheres. Gold non-tinning lands (Fig. 8) are probably a good compromise, but compatibility with the tinnable dot and the solder presents problems. The adhesion to the underlying gold land of most tinnable dots that were tried -- such as platinum, gold-platinum, and ternary alloys of noble metals -- was poor; the dots could be flicked out after they were fired. Silver bearing dots, such as silver-palladium, adhered reasonably well, but the ratio of 80 silver: 20 palladium did not tin very well in lead-tin solder -- apparently due to a change in the ratio when the silver is pulled out during firing to alloy with the gold in the underlying lands. However, a ratio of 85 silver:15 palladium tinned quite well, while higher silver ratios eroded too readily in the solder. In general, unless the solder is already saturated with silver, silver from the lands will tend to dissolve quite rapidly during tinning. The main problem with this silver-palladium and non-

75-77	BROWN GOLD POWDER
0-4	LEAD BOROSILICATE GLASS FRIT
0-3	COLLOIDAL SILICA
0-7	WETTING AGENT
15-25	METALLIC RESINATE – HANOVIA[*] 6566 MIXED WITH SILICON RESINATE

[*]HANOVIA DIV., ENGELHARD IND.

Fig. 8. Non-tinning gold paste formulation.

tinning gold combination is that lead-tin solder weakens the joint and tends to dissolve the soluble silver-gold rich alloy around the dot periphery. However, this can be minimized if tin is eliminated from the solder, since it is mainly the tin which erodes both the dot and the underlying gold land. If there is silver in the solder, such as silver-lead, only the dot seems to be eroded, but not the gold land. Thus, the metallurgical compatibility of this system, as mentioned before, obviously involves the functional solder system.

Although adhesion tests which involve pulling a reflowed metal rivet from a large dot of fired electrode do not necessarily represent the much smaller geometries of finger ends on circuits, a few adhesion tests were made on several tinnable lands fired on non-tinnable gold. The obviously weak joints of the non-silver bearing tinnable dots were shown, as well as respectable values for the superimposed silver-palladium dots. The results were quite erratic and not very reproducible; some typical values are listed in Table VI. With compatible metallurgical systems, however, the strength of the devices joined by means of dots can be equivalent to those joined to dammed sections or any of the other reflow chip joining methods.

Table VI. Some Adhesion Values - Dots.

Top tinnable land on gold non-tinning lands	Average tensile adhesion, psi
Au:Pt:Pd	1080
Au:Pt	rivets fell off
85Ag:15Pd	2443

Data by A. Wager - IBM CD

A metallurgically compatible dot method would offer the following advantages: high conductivity, minimal silver migration (silver does not readily form dendrites from a dot surrounded by non-silver bearing conductive material since the fields are strongest at the conductor edges), excellent tinnability, and slightly less critical registration of the two screened portions. In elaboration of the last two advantages: (1) the tinnable dots are not sterically shielded, as with dams, but instead are raised above the lands so that contact with the molten solder is excellent. Tinning yields on runs of several hundred parts approaching 100% were achieved when the proper metallurgical systems were used with this method, with much less care required than with dams; (2) since the dot shape is not dependent on registration, the land fingers can be extended beyond the normal chip pad positions so that minor misregistrations can be tolerated with some formats. The main limitation here is avoiding electrical shorting of the fingers with the conductive dot material if it seeps beyond the desired format during screening. If two parallel pad rows on the chips can be spaced further apart than the other two, greater registration tolerance which is not possible with dams can be achieved in that direction. Furthermore, for devices with relatively large pad spacings, the lands can be made quite large and the dots then have a wide latitude of positioning within the land area.

Chapter 7

SURVEY OF CHIP JOINING TECHNIQUES

  <u>Introduction</u>. The number of techniques for electrically interconnecting semiconductor devices has proliferated enormously during the last few years.[57,58] If all the possible perturbations of temperature heirarchies, materials, and different metallurgical processes are considered, several hundred variations are now discernible. This chapter attempts to provide a compilation, structure and review of those methods. Many of the methods are described and discussed, along with some of the materials and processing.

  A survey of this type always is subject to considerable limitation, and indeed in some aspects - futility. Completeness requires encyclopedism; incompleteness fosters challenge. Specific examples can seem self-serving; broad generalities can be unsupportable without such examples. Arguments valid for one manufacturer with certain facilities can be invalid for another. Despite these obvious limitations, it is still believed that a survey of this type can be useful in the decision-making process.

  There are several major considerations that should be weighed very carefully before individual processes are discussed. First, a distinction should be made between demonstrated reliability and postulations about reliability. Microelectronics has entered an era where the obvious can no longer really be trusted; simple, direct solutions to problems have shown an irritating manner of creating new, more subtle, even more aggravating problems than the ones they solved. This has become particularly annoying in assessing chip joining methods, since many of them are relatively new and have not accumulated sufficient reliability data to be truly predictable. It is comforting for example, to postulate that simple metallurgy systems should have high reliability - it is quite another thing for this to be proven, and indeed for it even to be true. Such simple metallurgical systems as gold-to-gold thermocompression bonded wires have been shown to be a function not of the metallurgical system, but more of the individual bonding process - which have been to be somewhat variable.[75,76] The mechanical aspects of microscopic joints require an assessment today which is considerably more sophisticated than the routine mechanical analyses of previous years.

A semantic problem that can interfere with an understanding of chip joining methods is the confusion of terminology. Since no standard glossaries are yet available, some nomenclature is laden with mutiple meaning, and the assessor of chip joining methods must be careful not to assign meanings to words which are not intended. We will try to define the context of the terminology as much as possible.

Another aspect of considerable importance is the relationship of the chip joining method to the other requirements of the module. Indeed, it is often other requirements that dictate the constraints upon the chip joining, such as temperature heirarchies, need for replacement, hermetic sealing, etc. It is therefore not unexpected that no single chip joining method is optimum for all fabricating conditions; chip joining methods become more or less optimum as the constraints around them change.

Survey Format. To reduce the complexity of existing chip joining methods to a convenient format, Figures 1 and 2 show a flow-chart type of arrangement. This format is an arbitrary one and is intended for general surveillance. The methods are separated into two main form factors - contact down (commonly called flip chip), and back joined, where the thermal path is directly through the back of the device onto a module surface or heat sink attached to the surface. This is an over-simplification, since techniques are available for providing both characteristics in the same fabrication scheme. For example, devices can be joined to lead frames, and then back joined to a module. This would then represent a flip chip becoming a back joined device.

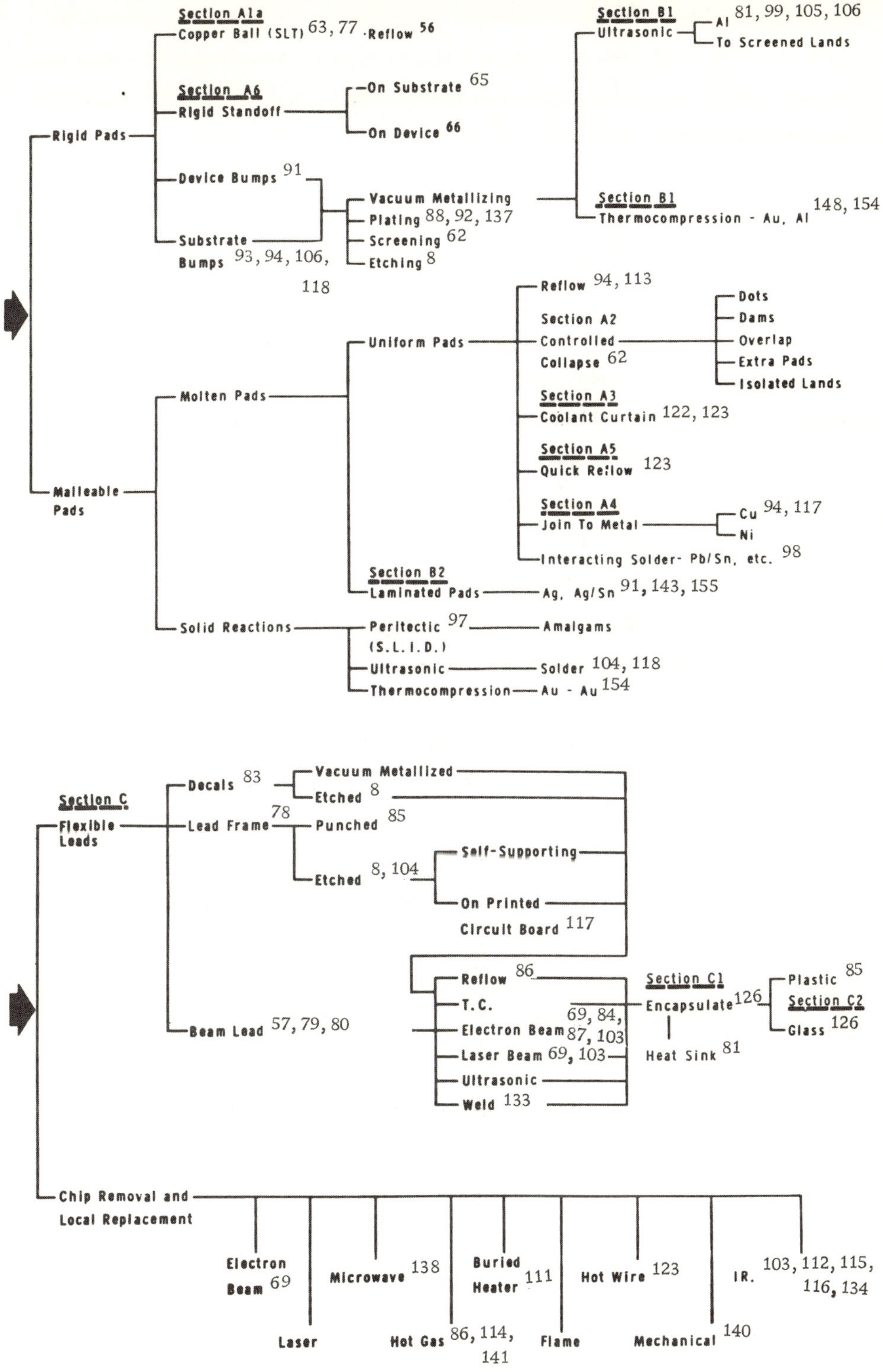

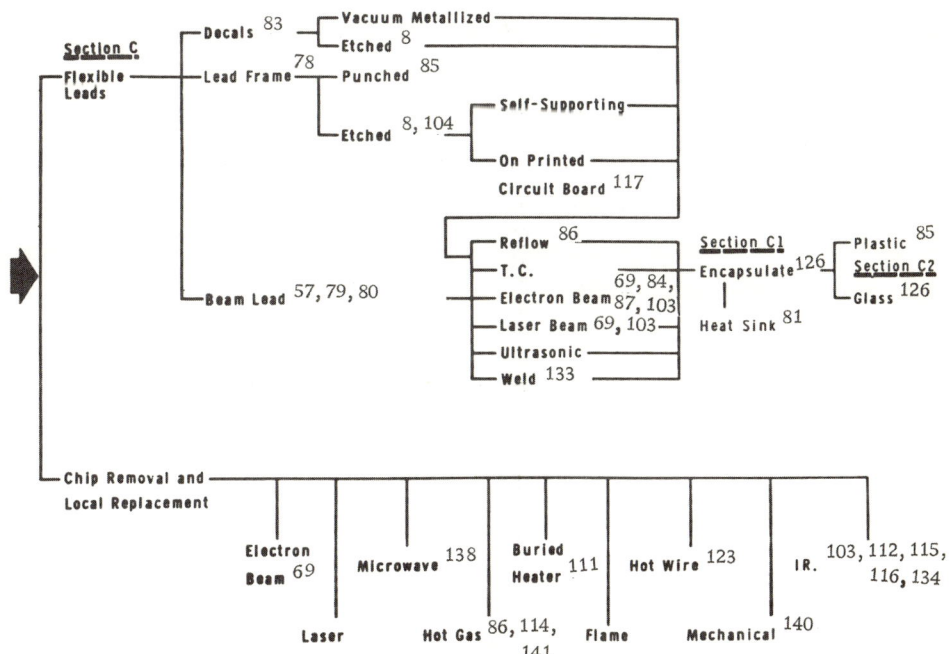

Fig. 1. Survey format, flip-chip configuration.

To permit a deeper understanding of each of these methods, appropriate literature references are placed with each procedure. Since the references are provided mainly for greater elaboration, no attempt has been made to define the references of greatest historical value or to include all the many papers on each topic.

An assessment of Fig. 1 (flip-chip configurations) shows that at least three major types of contacts have been suggested: rigid pads (such as the SLT copper ball),[77] malleable pads (such as the solder pads of the controlled collapse process or other reflow techniques), and bonds to more or less flexible carriers (such as decal leads or lead frames).[78] As integrated circuits become larger and more complex, there seems to be a greater emphasis on malleable pads and carriers than on rigid pads (except that ultrasonic bonding of aluminum-to-aluminum is still in active use). Beam leads occupy an interesting position, since with different metallurgies or different processes they also can be adaptable either to flip-chip or back joined configurations.[57] This technique has received considerable comment in the last few years.[79,80]

A perusal of Fig. 2 (back joined configurations) shows two alternative main approaches: the device pads can be joined first to some form of carrier and then have the back fastened down to the module for thermal dissipation, or alternatively the device back can be joined first to the module, and then the pads electrically connected. Furthermore, it is apparent that the device can be back joined either to the surface of the module[56] to a special heat sink on the module,[81] or imbedded in a cavity in the module.[82] The processes required for metallizing these various surfaces may differ.

Despite the arbitrary subdivision of the methods into front and back joining, many of the actual joining procedures - such as welding, reflow, etc. are pertinent to both styles. Similarly, many of the different types of conductors can be used in both contexts - such as decals, lead frames, and even ceramic modules.

Considerations. Many of the methods in these two charts will be discussed in some detail, but it is appropriate first to describe some aspects of similarity which relate to all of the configurations.

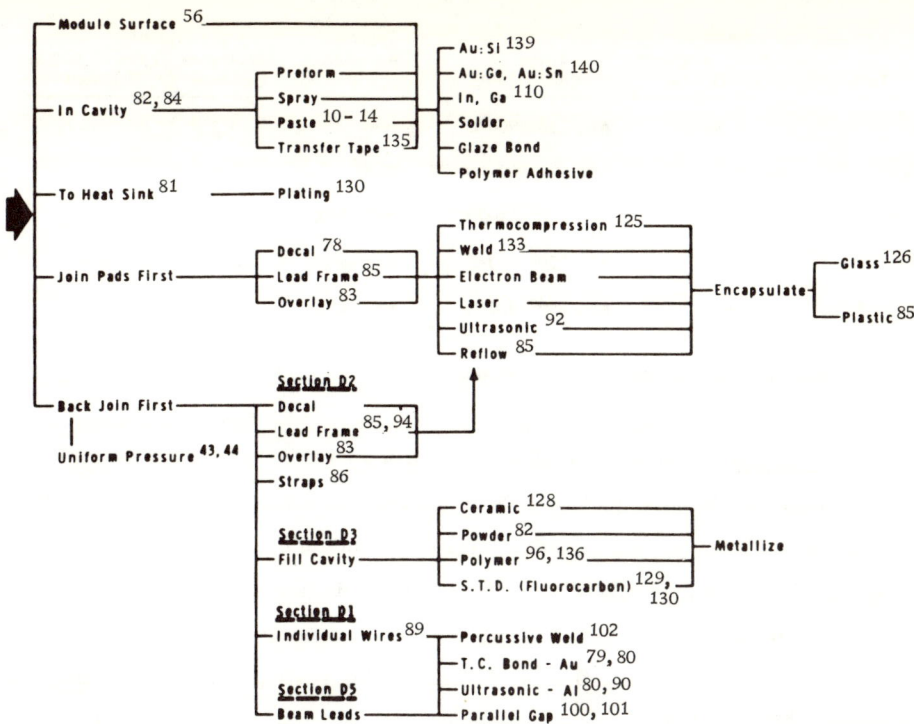

Fig. 2. Survey format, back joint configurations.

Conductor System. The first aspect to be considered is the conductor system to which the device is joined.

Screened Lands. [41,13,11] (see chapters 1-4) Screened lands commonly are limited to about 4 mil lines on 8 mil centers and fired heights in the order of about 1/2-2 mils. Conductivities are dictated by the metallurgy, paste transfer characteristics, screening process, paste solid loadings, and firing conditions. Figure 3 summarizes typical paste systems and permits both a comparison and an estimation of limitations. Each paste was screened through 200 mesh with 1 mil emulsion thickness. Screening operations are considered to be of relatively low cost compared to other types of conductor fabrication schemes.

Paste	Metal or alloy resistivity $\mu - \Omega - cm$	Firing Conditions	Typical Adhesion P.S.I.	Typical Resistance, Ω/in./0.010 in. Untinned	Tinned
Ag	1.6	Air, 750°-900°C	----	0.2	---
Cu	1.7	Reducing 900-1000°C	2500	0.3	---
80 Ag:20 Pd	10	Air, 750°-1100°C	2500-3000	1.5	1
(Au:Pt)	---	Air, 750°-850°C	2500-4500	6.5	0.8
(Au:Pt:Pd)	---	Air, 750°-850°C	4000-4500	7-11	1.4

Fig. 3. Summary of typical paste systems.

Thin Films. Vacuum metallized thin films have been used for years in chip joining. These have generally taken the form of aluminum, gold, or copper with appropriate underlying adhesive layers of titanium or chrome and sometimes barrier metals such as nickel or moly to avoid interactions. Thin films can be used on rigid substrates such as glass or ceramic, and have been considered for use on relatively flexible films. An interesting variation for handling relatively thin metal conductors involves decals, which provide support to fragile conductor fingers during chip joining, but from which the plastic backing can be removed afterwards.[78,83] Such decals can be used for contacting devices which are back joined to module surfaces or cavities.[84] Two methods are plausible: either join the chip first to the decal on its carrier and then join the chip back down into the module cavity (Fig. 4a) or join the chip into the cavity first and then attach the decal leads (Fig. 4b). With the former technique,

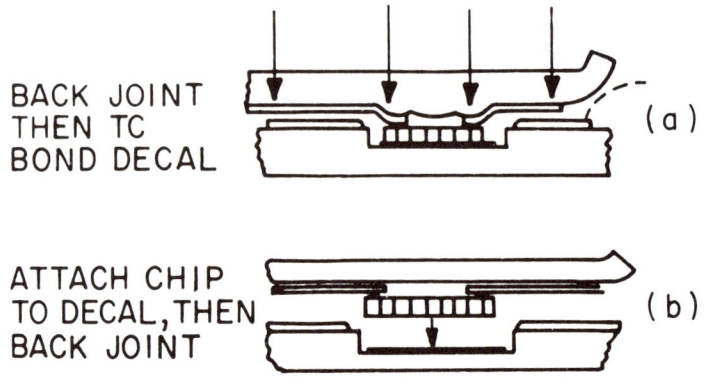

Fig. 4. Decal back joining.

registrations may be simpler since device pads can be aligned with the decal
lands and in effect then become a beam lead structure. If the latter process is
used, a triple registration involving the chip, decal, and module lands must be
accomplished. If decals of very small area are utilized, the material costs may
not differ much from the cost of the entire screening process. Conductivities
approximate bulk metal values.

Self-Supporting Lead Etched Frames. Lead frames with small dimensions
can be etched through photographic resists from thin sheet metal stock.
Punchings[85] which might be cheaper, are more limited in the attainable minimum
line width, require expensive dies, and therefore are not readily subject to
design changes. Punchings have been made from 4 mil thick Kovar with fingers
about 8 mils on 16 mil centers (Fig. 5). Production dies for such parts might
cost as much as $15-20,000. If lead frames are etched from both sides, even
finer lines can be obtained - such as 4 mil lines on 8 mil centers. Platings can
be applied to modify the surfaces as required: such metals as gold, nickel,
copper or solder[86] can be plated on such frames as Copper, Kovar or molybdenum.
Indeed, the poor electrical conductivity of Kovar normally requires improvement
with such platings. Obviously the chioce of lead metal may be restricted by
subsequent processing conditions; molybdenum, for example, may oxidize
drastically during binder burn-off if a glass paste or slip preform were fired in

Fig. 5. Lead frame.

air to bond the frame to a support carrier. All metal surfaces other than gold generally require cleaning or etching before chip joining if solder reflow techniques are used, but perhaps not for thermocompression bonds or ultrasonic bonds. The fingers of lead frames are often unsupported, and can give handling difficulty. However, such lead frames can be bonded to supports - either removable or non-removable - for extra mechanical control. In the latter case, such supports in effect convert lead frames into decal structures.

Beam Leads.[79,80] The metallization leading from the device can be formed with the device itself, generally by plating, and can be connected to any of the previously mentioned types of metallizings - screened lands, thin films,[87] lead frames.[88] The most common type of beam lead has been gold, but some interest has been expressed in aluminum. Barrier metals such as platinum are used to prevent layer interactions. No major technical reason is seen why other metallurgical lead systems - such as copper or other solderable materials - could not also be considered. A major consideration with beam leads is the expected relatively high device cost.[57]

Wires. Perhaps the most common method of chip joining to date has been thermocompression or ultrasonic bonding of individual wires to back-joined integrated circuits.[89] The metallurgy almost universally has been either gold (most common) or aluminum. Mixed Au-Al joints have provided reliability problems with intermetallic compounds of Au and Al (the vari-colored "plagues").[90]

Pads or Bumps. A wide variety of techniques and materials have been utilized for creating the bumps which interconnect the flip chip and module. The SLT process reflows nickel plated copper spheres about 5 mils in diameter onto vacuum metallized areas of lead/tin solder on the wafer.[62] The controlled collapse solder pads are vacuum metallized larger than the ball limiting pad metallurgy on the wafer and are then reflowed in an inert or reducing atmosphere to form hemispherical pads. The Hughes flip-chip is constructed by vacuum metallizing the various laminations of silver and silver/tin.[91] It is not outrageous to consider depositing even several mils of low melting, high volatility metals such as the lead-tin-silver type.[64] However, less volatile metals can

134

take excessive time for deposition. Thus, electroplating and electroless plating can be used for building up higher melting metals such as gold.[92] Typically, for example, the beam leads and even some of the gold thermocompression bonding pads are built up by plating processes.[88] Other relatively crude techniques such as screening or dipping have been suggested for putting on bumps, but these have not apparently received wide usage because of their lack of uniformity.[62] If it is desired to put the bumps on the substrate, methods similar to the above can be used.[93] Very often, a bonding layer of chromium is topped by a vacuum metallized layer of aluminum, copper or gold. But the bumps on the modules can also be made by etching away adjacent portions of the lands, leaving a raised bump section,[94] or by impressing or casting the bumps. Similarly, raised surfaces can be obtained by screening an extra layer of electrode paste on the area to be raised. The generally thicker land sections on the substrates are more amenable to etching and forming techniques than the relatively fragile device surfaces. Many of these techniques have been used to fabricate structures economically, reliably, and efficiently.[95]

If bumps are applied by vacuum metallizing techniques, the masks must be changed or cleaned periodically, since the build-up of deposited material on the mask tends to narrow the size of the pedestal. Reductions of about 20% of the diameter have been observed in four depositions of aluminum when the masks were not cleaned. A taper can be created on the pads by spacing the mask from the device surface; this is desirable since the taper can permit squashing of the thinner tip sections of the longer pads to reduce problems with planarity.[76]

<u>Chip Joining Process Considerations</u>. The second aspect to be briefly considered is the chip joining process itself.

<u>Soldering</u>. Reflow processes (and some ultrasonic processes) involve mainly the alloys of lead and tin, although silver and alloys of indium or gallium have been mentioned. However, if compatibility with higher temperature processes such as hermetic sealing is required, such low melting solders could be replaced with higher melting ones.[96] Higher melting solders are indeed available, but higher temperature solder fluxes would also be involved, possibly introducing potential corrosion worries due to activated fluxes or flux residues.

The temperature heirarchies of module fabrication are extremely important - in fact, they severely limit some methods. The upper temperature limit is generally defined by the device pad metallurgy - normally somewhere about 450°C. The lower limit is defined by test and usage conditions, with a sizeable safety factor - probably in the order of about 200°C. The maximum number of temperature processes between these limits is defined by the permissible separation of processing ranges - which might be as low as 25°C, but preferably 40-50°C for latitude in manufacturing. Most molten metal processes (as soldering, for example) must be performed at temperatures considerably higher than the liquidus. Thus, the controlled collapse joint is made at about 340°C, even though the 10 tin/90 lead solder solidus is about 260°C - which subsequent processes should not exceed by very much. As further illustration, consider the following process: a device is back joined with gold-silicon eutectic at about 400°C; that joint should not now be taken above about 370°C. If leads are joined to that device with 10 tin/90 lead solder at 340°C (which is only 30°C from the lower limit of the gold silicon), the resulting joint should now not be taken above about 280°C. Thus any other processes such as sealing or joining to the next level should not exceed about 240°C. This obviously can be a severe restriction. Metals with more limited "slushy" regions between the liquidus and solidus require less range.

Peritectic Systems (Solid-Liquid Interdiffusion).[97] Solid-liquid interdiffusion bonds have been considered for joining devices and making seals. With this technique, two applicable solid metals or alloys are pressed together at an elevated temperature for an extended time and a diffusion joint results with a considerably higher liquidus than the diffusion temperature. This technique might be usable with decals or lead frames to permit subsequent higher temperature processes on chip carriers, but it does not seem to be beneficial for chip replacement due to the very high replacement process temperature and to the change in the remaining metallurgy when the device is removed. It also requires rather extensive process time, coupled with the need for continuous pressure of the chip against the substrate. Non-peritectic systems which form higher melting alloys than the processing temperature by interdiffusion have also been demonstrated; one such system uses lead/tin solder.[98]

Welding Methods. Welding techniques offer an alternative to soldering and may require different materials. Thermocompression bonding might be considered a welding process, but it is restricted to ductile, malleable metals, and usually is made either with gold wire of very small dimension or with gold beam leads. Ultrasonic bonding has been successful with aluminum wires and pads, and even with flip chip configurations - particularly if the pads on the device or substrate are tapered to reduce problems of planarity.[99] Other welding techniques include microparallel gap welding[56,100,101], percussive arc welding[102] of small wires, etc. These have been the subject of studies, but little industrial usage for chip joining.

Thermocompression Bonds.[90] Thermocompression bonds - with the exception of recent work with multiple bonding of beam leads[87] - have typically been restricted to the common ball, stitch, or wedge bonds which are single lead operations, with one piece of thin gold wire joined at a time. Ball bonds probably have the best strength; wedge bonds tend to weaken and deform the wire. For single wire processes, estimates of 0.25 to 1 second per bond have been made. Unlike soldering, but like all welding type processes, both temperature and pressure require control. Joint strengths have ranged to 10 grams per joint, but even this relatively low strength is probably reliable if uniformity can be maintained. Good flexibility and aging characteristics are certainly obtainable with ribbon-like gold to gold bonds, but there are indications that control of the strength and reliability may be difficult, as operators or equipment are changed. Chip replacement also has problems, since every wire has to be considered separately. Certainly wires can be pulled off and new wires joined on an individual basis, but this is considerably more complex than the replacement of a flip-chip device in one operation. Wire bonding is particularly advantageous for modules which do not require chip replacement, for prototypes or small runs where tooling would be prohibitive, or as a repair technique for individual broken decal or lead frame connections.

Lasers and Electron Beam.[56,69,103] Mechanically sound joints can be made to devices with electron beams or lasers, where true welds are obtained with relatively high strength and minimum lead distortions. Physical contact of the lands and pads must be maintained while the programmed beam rapidly fuses each joint in sequence. Electron beams can be very rapid, make enormous numbers of clean bonds, and be used for hermetic sealing at low temperatures as well. Laboratory equipment can be extremely expensive, although limited parameter equipment for production usually is significantly cheaper. Since many modules could be welded with each pump-down in an electron beam system, a high degree of automation is suggested to make the process economical. Lasers do not require the vacuum system of the electron beam, but still may need inert atmospheres to prevent oxidation.

The main difficulty with both procedures is the delicate control of the high intensity beam; too little energy makes a weak bond - too much energy might destroy the device. However, neither of these two methods should be ignored, since they are both techniques which are being rapidly sophisticated.

Ultrasonic Bonding. For flip chip configurations, the chip is clamped against the substrate metallizing and ultrasonic vibration is utilized to bond the pads and lands. Although the predominant usage has been aluminum-to-aluminum, other metals are also applicable and solders have been demonstrated - such as lead/tin, silver/tin, gold/tin, and cadmium.[104] Copper projections have also been used, either evaporated or plated. In one group of experiments, low yields were achieved with 40 pad devices, while those with 14 pads achieved high yields.[105] Replacement of ultrasonically bonded chips normally requires breaking off the chip at room temperature and then re-bonding another. However, if the best ultrasonic bonds are obtained, the joints are so strong that the lands tend to pull off. Thus, the ultrasonic operation is normally modified to produce only a moderately strong bond so removal is more realistic.[95] Tensile strengths of 20-40 gms/pad are obtained.[76,106]

Methods of Back Bonding. Three types of back joints are common: solder reflow, eutectic bonding, and adhesive. The first two can be replaced and provide somewhat better thermal dissipation than the last. The main problems with these joints relate to device alignment, and the need to create a stress-free joint that will not rupture on thermal cycling. Solders tend to contain more air gaps in the bonding layer, and particularly with very large devices, both solders and gold/silicon or Ge/Si eutectics can be subject to cracking on thermal cycling. Since gold/silicon bonds are commonly made at about 400°C, they are more compatible with subsequent relatively high temperature processes – but they cannot be used if lower temperature solder reflow techniques were previously used to join the leads to the device before the back bond. On the other hand, the lower temperature lead/tin back joint permits prior bonds with higher melting solders, but not subsequently high temperature processes. Methods for both evenly distributing the joining pressure and scrubbing have been proposed.[107,108]

The metallurgy on the back of the device of course must be adapted to the bonding method in either case, typically gold (for gold-silicon bonding) and chrome-copper-gold, as an example, for solder reflow. System thermal conductivity values are probably not apprecibly different due to the very thin sections involved. Epoxies and other polymers are difficult to remove if thermosetting, and are less thermally conductive unless heavily filled – which can reduce the joint strength. Glazes are also less thermally conductive but can provide a better thermal expansion match to the device and substrate. Back bonds are often made to plated metal surfaces for optimized thermal paths.[109] Ductile metals can be used (such as In)[110] as well as the more brittle Au intermetallics or alloys.

General Considerations. The next aspect to be discussed involves general considerations.

Methods of Repair and Replacement. The very high cost of extremely complex multichip modules will probably not only permit, but indeed demand, that broken conductors be repairable, preferably before devices are committed. Certainly the economics of the degree of permissible repair has to be assessed, and redundancy of some sites on very complex modules will help reduce such problems. As typical examples of repair, broken screened lines can be repaired with paste touch-ups applied by fine brushes which are refired, or by applications of preforms and decals. Fine wires can be reflowed or welded to heal gaps. Repair of thin film circuits might be attainable with localized plating techniques, but care should be taken to avoid stress buildups. Formats utilizing solder covered lands are probably more readily repairable than others.

Repair of individual leads to back mounted devices is not considered difficult, whether they were reflowed or welded. Individual faulty leads can be removed with the aid of localized heating if necessary, and replaced with wire flying leads. Certainly TC bonded leads can be replaced at will; even if weld residues are left on the lands, there should be sufficient uncontaminated land area remaining to which replacement joints can be made.

Chip replacement is considerably more complex with back mounted devices where not only the back joint must be released, but also all the leads. Lead fragments remaining on the module lands may be detrimental to joining the replacement chip. If solder is used, it would be preferable for the pad-lead joint to have a higher melting point than the lead-land joint so that a controlled replacement temperature would melt only the latter and permit the leads to remain attached to the device as it is removed (such as by vacuum). This is consistent with the other benefits of joining the device pads down to the leads (either decal or lead frame) before the back joint, such as easier registration and handling.

As previously mentioned, both solder and gold/silicon eutectic back joints are replaceable - the latter at a higher temperature. The need for not disturbing adjacent good devices is obvious, so that configurations and heating techniques, (including thermally isolated heat sink regions, buried heaters[111], usage of infrared[112,113], hot gas[114], lasers, hot wires, etc.) must be carefully considered. Indeed this is one of the critical factors now facing the design of complex multichip surfaces. (Incidentally, infrared has multiple advantages in being capable of acting both as a localized focused heat source and for alignment viewing since silicon is transparent to certain IR wavelengths.)[115]

There is little difficulty in replacing reflowed flip chips - multiple replacements can be made rather conveniently.[116,117] A consideration here is to restrict the heating zone to leave adjacent devices undisturbed. Since no back joint is involved (except perhaps for a heat sink on the back of the device if desired), no temperature heirarchy is required.

Some chip carrier schemes require that the whole carrier be replaced - such as with an encapsulated lead frame and chip. Others do not permit replacement readily at all - such as the STD process described later. If thin film lands are strongly adherent, chips joined either by TC bonding or ultrasonic bonding can be sheared off and replaced by the same process. However, there are indications that at times such processes lead to the removal of the thin films themselves, requiring compromising the strength of the joint to prevent this from occurring.[118]

<u>Line Widths and Dimensions</u>. A rational dividing line for screening is about 4 mils line width. With careful control, screening can produce such lines on 8 mil centers. Closer dimensions are not considered usable for large quantity production although they have been demonstrated in the laboratory, for several reasons: difficulty in making etched masks (emulsion screens are not realistic in this range); severe clogging problems; drastic reductions in transfer thickness (and thus high electrical resistance); high susceptibility to solder erosion. Etching processes are limited by the depth-width ratio. Commonly, metal etching results in undercutting of 1/2 mil to 1 mil per mil of depth. For a nominal 4 mil wide finger etched from 2 mil stock, this undercutting could be from 1 mil to 2 mils - a large proportion of the total width. For smaller lines, the undercutting becomes prohibitive unless the stock thickness is drastically reduced. Unsupported lead frames have low handling strength if less than about 2 mils thick; thus such frames are probably restricted to about 4 mils width. However, thinner supported decals might attain 2 mils if etched and almost certainly if vacuum metallized through a mask (which probably is a bit more costly).

A realistic assessment of other printing processes (such as lithography, flexography, intaglio, xerography, etc.) indicates that none of these are readily applicable to attaining uniform lines adaptable to circuitry which would be narrower or better than those attained with screening, vacuum metallizing, or photo-resist etching. Beyond about 4 mils, photoresist techniques will probably have to be used, whether for direct etching of decals or metallized surfaces, or for producing the masks required for vacuum metallizing. (Note: Electron beam engraving has not been evaluated here.) It should be recognized that fine line processes will be costly - particularly due to potential yield and repair problems - and thus should be used only when necessary. In addition, the high electrical resistance of very fine lines may provide practical limits to circuit design.

Cost assessments can be very complex in such matters. The device designer might be able to save considerable real estate by decreasing the pitch of the device pads. However, this saving might be sacrificed if the module designer has to increase his costs disproportionately to adapt to those spacings or if yields are drastically reduced. Furthermore, aspects of reliability must be considered, since smaller joints can be weaker and less controllable. However, very high circuit speeds or other requirements can certainly overcome such arguments in individual cases.

Thermal Considerations. Most integrated circuits probably do not have isothermal surfaces, but instead consist of a series of small hot spots at the junctions.[119] Good thermal conductors (metal heat sinks) can be fastened to the device to spread out these hot spots and diminish their effects; certainly the maximum number of thermal paths should be utilized. Very high powered devices will probably not be able to dissipate enough heat through flip chip pads, but will require extra help - even if the substrate were modified to include a number of high thermal conductivity sections. In such cases, the heat must probably ultimately be removed from high power density areas by some form of liquid cooling. For less dense areas, air circulation past thin heat sinks may be sufficient.

Multi-chip modules can be small self-contained thermal packages using internal immersion cooling[120], Fig. 6. However, if it is true that a realistic limitation for air cooling is about 4 watts/sq. inch, such self-contained units might be limited to relatively low power densities anyway. A quick calculation based on such a package indicated that even 25 watts/sq. inch might be excessive. Such considerations require decisions as to whether to pump coolant through external heat sinks or through the module itself, whether to have the coolant in direct contact with the devices, etc. Direct contact of the coolant with the device is preferable if no mechanical damage or corrosion results. However this may pose mechanical interconnect problems at the machine level.

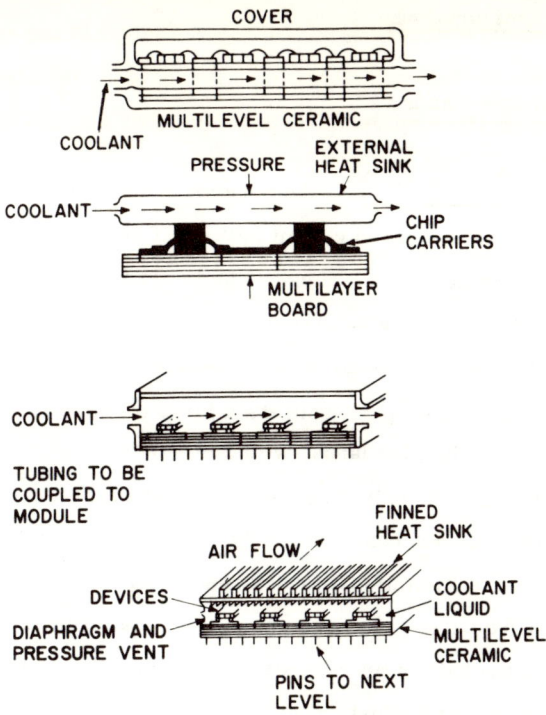

Fig. 6. Several types of liquid-cooled modules.

It would seem at this stage that most configurations can be used for low power densities - permitting the usage of relatively low cost flip chip techniques - while medium power densities may require some form of back joint so that heat can be dissipated through the module. Very high power densities ultimately will be controllable only by some form of liquid coolant, where again the configuration of the chip is less important. In the latter case, chips with contacts down and the proper heat sinks on their back can be just as effective as back joined devices.

<u>Description of Typical Chip Joining Processes</u>. We will now describe some of the many possibilities of chip joining in more detail here, together with some of their advantages and disadvantages. As previously mentioned, each technique can be varied considerably, but we must restrict ourselves to a typical fabrication scheme for each major process. The individual methods chosen for this discussion were picked either because they illustrate an interesting facet or have generated particular industrial excitement.

A. Flip-Chip Configurations

Reflow to Solderable Lands

1. <u>SLT Process</u>[29,61] (Fig. 7). The ceramic module is fabricated by screening conductor paste in the desired circuit pattern, firing it, and then tinning it. The tinned acceptor regions on the fingers can be flattened or dimpled to receive the copper contacts on the device. Typical metallurgy systems are either gold platinum or silver palladium lands, lead/tin solder, and nickel plated copper balls joined to the chrome-copper-gold metallurgy of the device.[62] The device balls are placed on the dimpled lands, with a sticky layer of rosin flux to hold them together, and the composite is passed through a reflow furnace to provide a solder joint. This method has been highly successful, automatable, and reliable[70], - but it has limitations for devices requiring more than the relatively few pads of the SLT devices. Furthermore, the metallurgical complexity and unusual aspects of the balling procedures have apparently prevented this method from receiving widespread industrial usage.

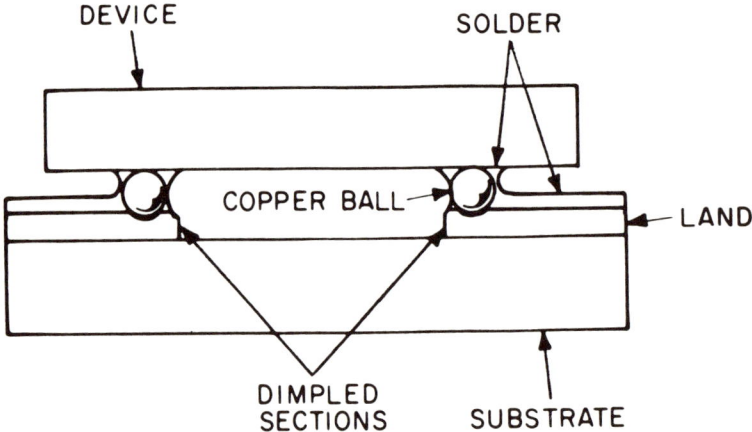

Fig. 7. SLT device.

2. <u>Controlled Collapse</u>. (See Chapter 5) (Fig. 8). This method is very similar to the SLT method, except that the solid ball is replaced with a malleable pad of vacuum metallized lead/tin solder. Furthermore, the tinnable chip joining sites on the lands are surrounded by non-tinnable barriers so that when the solder on the lands and device pads melt and merge, surface tension holds

the chip suspended above the substrate. This method is simpler metallurgically, retains all the other advantages of the SLT method, has an even more reliable ductile joint[67] and is extrapolatable to more complex devices requiring many pads and larger size. It does not require passivating the device edges, as early solder pad work suggests.[121]

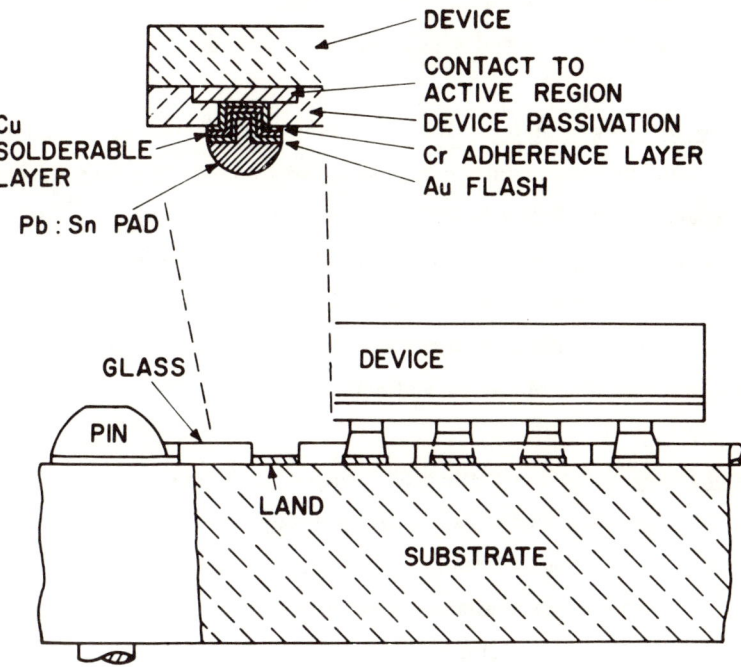

Fig. 8a. One form of the controlled collapse method.

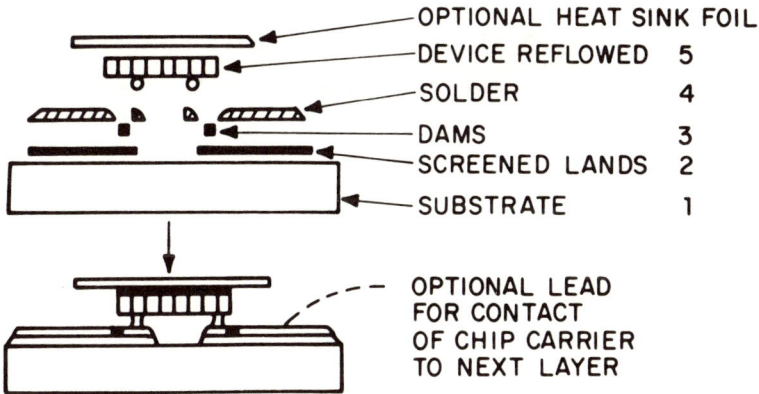

Fig. 8b. Fabrication of typical controlled collapse module.

3. **Nitrogen Curtain.** Another method of preventing solder pad devices from collapsing involves a coolant curtain of inert gas beyond the periphery of the device.[122] This also limits the molten zone to the pad section. This method probably has limitations in the degree of control required for attaining uniform device heights, and requires somewhat specialized joining equipment - which may not be a severe limitation. The inert curtain may preclude the need for flux, and stress can be relieved with infrared reheating.

4. **Reflow Directly to Metal.** Reflowed joints can be made to certain unsoldered metals, and if carefully controlled, the solder on the device pads does not flow out and wet the entire land, but remains restricted to the joining zone. This may be due to slight oxide contamination on the bare metal surfaces. Metals which have been successfully used are nickel and copper.[94] This process must be carefully controlled in terms of the surface cleanliness, the type of flux used, and the balance between cold joints and collapsing.

5. **Rapid Pulse.** Another method for preventing the collapse of reflowed devices is to contact the device with a probe through which a current can be passed rapidly, providing an instantaneous heating of only the device and pad sections.[123] Equipment is available commercially and it is indeed a useful process, particularly for minimizing the formation of intermetallics. The device-to-substrate distance can be maintained if the back of the device is held against the heating probe by means of a vacuum - defining the clearance.[122,124] Since the chip joint is made at the time of placement, the machine is tied up for the full cycle - whereas methods A1 and A2 use the station for placement only, which helps to avoid a bottleneck. Such a consideration is mainly applicable to large scale operations.

6. **Standoffs.** A further means of preventing solder pad devices from collapsing involves the use of standoffs - either on the substrate or on the chip,[66] or even as temporary removable standoffs (Fig. 9). This is an interesting approach, but it does not seem to be used much. The standoffs can provide extra mechanical support and even aid the thermal characteristics if they are designed properly.

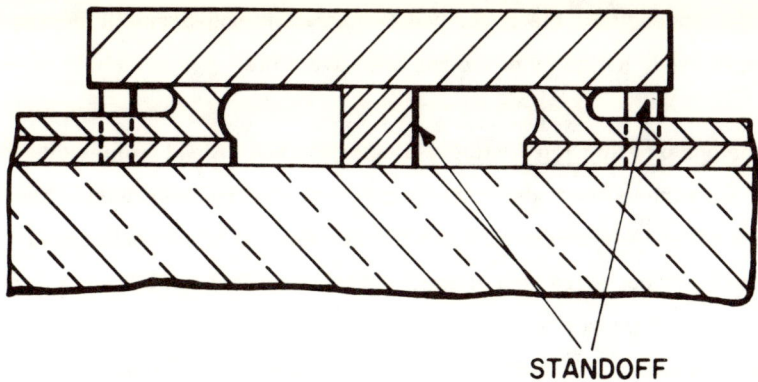

Fig. 9. Standoff, evaporated on device, preventing collapse.

B. <u>Other Flip-Chip Schemes</u>.

1. <u>Ultrasonic and Thermocompression Flip-Chip Bonds</u>. Both ultrasonic bonds and thermocompression bonds have been used for joining devices in flip chip style. With these techniques, the bumps can be either on the chips or on the substrate. For ultrasonic bonds in particular, both the lands and device pads must be reasonably planar, or the ultrasonic scrubbing action will pivot the chip. TC bonding tends to be a little more lenient in this respect, since the pads are crushed down and overcome some lack of planarity. The thermocompression bond requires a relatively high temperature of processing; the ultrasonic bond is very interesting in requiring no high temperature at all. Both methods - as in method A5 above - require that the joining and placement be made with the same piece of equipment. Beam leads are typically formed with multiple TC bonds.

2. <u>Laminated Pad</u>. An interesting variant which has combined both ultrasonic bonding and reflow bonding is the use of a relatively high temperature pad coated at the tip with a lower melting alloy. A typical example would be a silver pad, with a silver/tin alloy layer at the bottom.[91] The bond can be made ultrasonically, with the silver section preventing chip collapse to the substrate. Similarly, at a controlled elevated temperature, the lower melting layer can make a reflowed joint, with positive support from the higher melting portion of the stud. This method is in commercial usage and has received considerable interest. Specula-

tion about the tendency for the silver pad to migrate in non-hermetic packages has been expressed, but no evidence has been available.

3. <u>Extensions of Flip Chip Joining</u>. Devices joined to lead frames or decals by many of the various methods - including reflow, thermocompression bonding,[125] ultrasonic bonding,[92] microparallel gap bonding etc., can be encapsulated (either with or without extra heat sinks), and be made into a relatively inexpensive package. With proper tooling, such methods can be highly automated, and can be quite efficient,[83,85] and heat sinks can be also encapsulated into the package as required. The main considerations here are the proper mechanical configurations to avoid excessive strains at the frame-pad joint.[81] Furthermore, such devices should be passivated to avoid the need for a hermetic seal - which the common encapsulants other than glass are incapable of providing. This should not be restrictive; the industry is generally tending toward passivated devices, and it is probable that this will indeed be the path of the future. It is becoming more widely realized that the hermetic package is an unnecessary cost burden for most fabrications. An important contrary motivating factor in behalf of hermetically sealed packages has been some military requirements.

4. <u>Encapsulated Lead Frame</u>. A particularly interesting method of this type uses etched copper lands on Mylar or polyimide film to which the device is bonded face down. Encapsulant is placed over the entire device and land section, and the Mylar is subsequently removed (Fig. 10). The bumps in this method can be either on the device or on the etched copper.[83]

5. <u>Glass Encapsulation</u>. Another process encapsulates the chip and lead frame section in glass, providing a handleable high temperature package of very low cost. A wide variety of joining schemes were demonstrated, such as thermo-compression bonding, resistance welding, ultrasonic welding, and brazing and soldering. However, thermo-compression bonding of Kovar leads to aluminum pads was preferred, and some work on multiple bonding was done with 14 pad devices. After the chip is joined to the Kovar foil stamped lead frame, the composite is insulated with pyrolytic silicon oxide and then the device is sandwiched between glazed covers which are heated to flow the glass around the chip

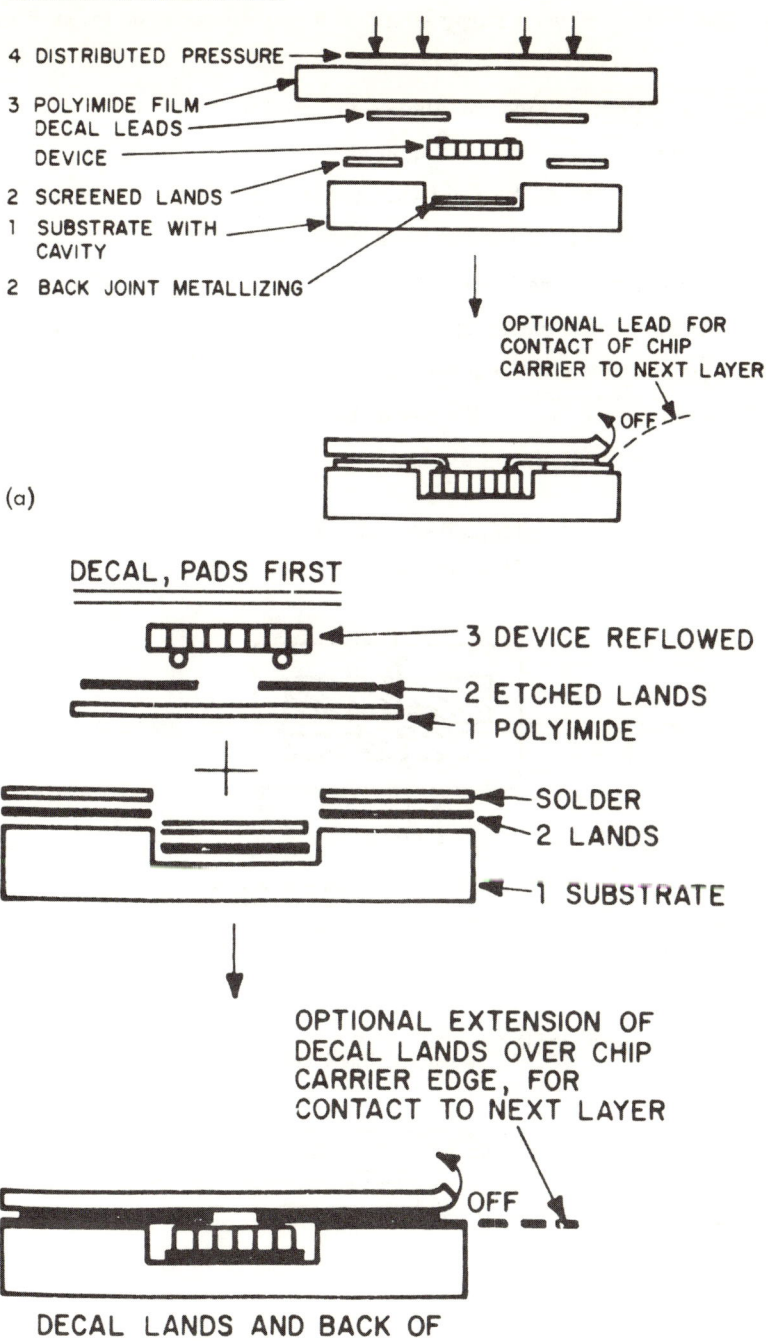

Fig. 10. Typical decal processes.

and frame section.[126] Considering the simplicity and obvious potential advantages of this method, it is somewhat surprising that it has not received greater emphasis.

6. <u>Other Variations</u>. Chips can also be mounted in the center of lead frames and then completely encapsulated by liquid hardenable resin to maintain the joint integrity[85] (Fig. 11). Another recent variation in flip chip joining consists of inserting the device into a cavity of a very thin substrate to contact printed circuit lands which overhang the front of the cavity. After the chip is joined to these overhanging flexible leads, the back of the chip can be joined to a heat sink metal on the opposite side of the substrate[127] (Fig. 12). Controlled collapse joining can be performed on lead frames if the solderable fingers are delineated by stop-offs such as on aluminum plating.

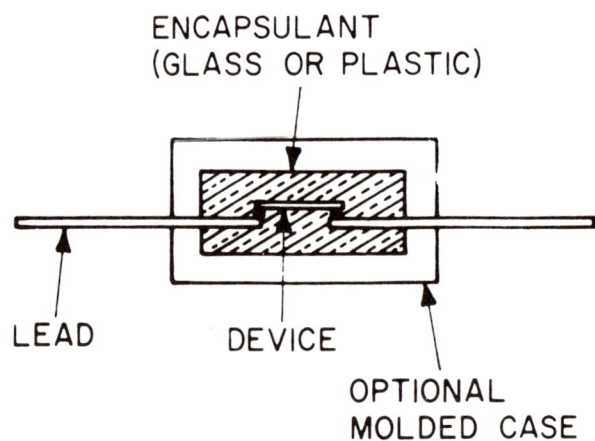

Fig. 11. Typical encapsulated lead frame process.

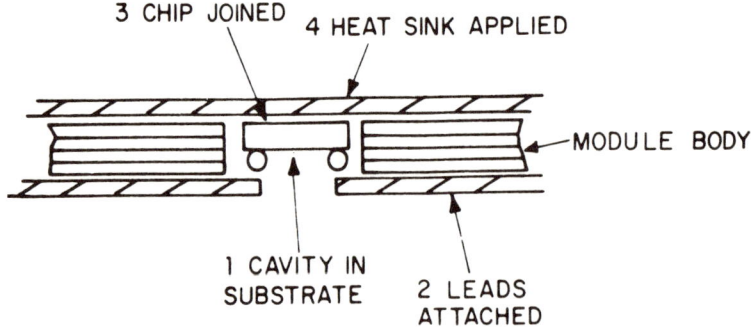

Fig. 12. Overhanging leads.

C. Back Joined Configurations.

1. <u>Wire TC Bonding</u>. Perhaps the most typical chip joining method in industry is the gold/silicon eutectic bond of the device back down to either a header or metallized surface on the substrate, with wire bonds to individual pads of the device (Fig. 13). While this obviously will be useful for many years for small orders and prototype models, it is not the preferable way for large scale production.

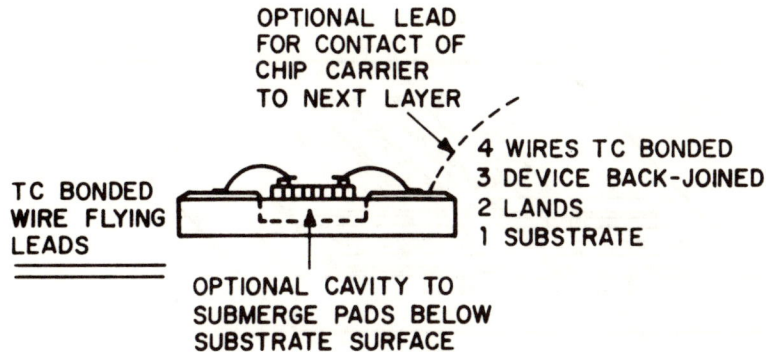

Fig. 13. Typical wire bonding format.

2. <u>The Use of Decals</u>. In this process, the chip is joined back down in a cavity of the substrate and interconnected to the substrate lands by conductors on a removable decal backing.[59,84] The backing is typically Mylar[7] or high temperature polyimide. As previously mentioned, the device may be joined to the decal first and then back joined if desired - and this is considered to be the preferred method. This technique can also be highly automated, and many kinds of joints can be utilized, including reflow, ultrasonic, TC bonds, etc. A factor to be considered here - as with lead frames[126] - is the need for making sure the decal conductors do not touch the edge of the chip, creating an electrical short. The flexibility of the very thin film conductors should be highly advantageous here - as with some formats of beam leads - to avoid stresses during thermal cycling.

3. <u>Filling the Moat</u>. Techniques (Fig. 14) have been demonstrated whereby the device is back joined to the bottom of a cavity, and the space between the chip and the cavity wall is filled with an insulator material - such as plastic

polymers or powdered inorganic material.[82,128] Interconnections can now be either vacuum metallized to make contact from the substrate lands to the device pads, or decals or overlay circuitry (where the plastic carrier is not removed) can be utilized. In general, these methods have not received wide acceptance primarily due to the mismatch of coefficients of expansion of the various materials and the subsequent rupturing of bonds during thermal cycling. Furthermore, the need to avoid the open edge of the chip provides additional difficulties.

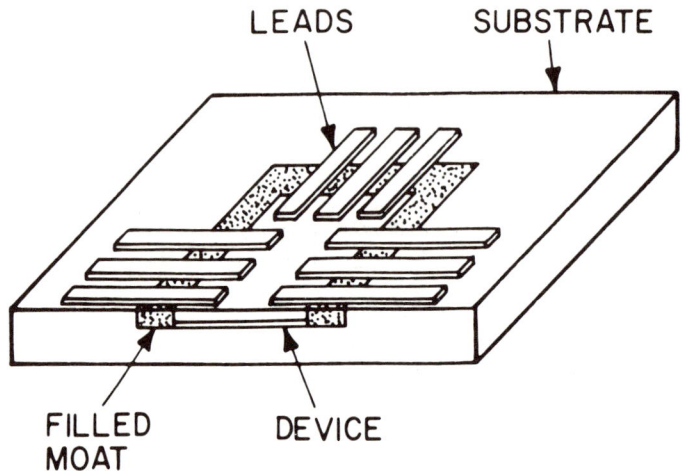

Fig. 14. Filling the moat and evaporating leads.

4. **STD Process.**[129,130] With this method the chip is not placed in a cavity, but the cavity is built around the chip by pressing warm plastic material around the device, which holds it in place. As with the cavity method, lands can then be evaporated across the surface of the plastic and through etched holes to make contact to the chip pads. This method might also involve problems due to potential thermal expansion mismatches of the materials, difficulty of chip replacement, and relatively poor thermal paths. Of course, with this method, as with any of the others, the hindsight of history will be necessary to determine whether these potential problems are indeed limiting. (See the next chapter for further information).

5. <u>Beam Leads</u>. As previously suggested, beam leads can be used in many different configurations, so that it is difficult to asses only one aspect. Because the predominant usage has involved thermocompression bonding all the leads at once, this will be the model to be used here.[88] A wide variety of tooling has been suggested for this purpose, some involving disposable deformable[87] anvils.[131] This is an area of particularly active development.[132]

<u>Conclusions</u>. This discussion has shown how difficult it is to be dogmatic about any single chip joining method. There are numerous factors - particularly related to cost, reliability, producibility, and circuit requirements that vary considerably for each manufacturer. There is considerable insight to be gained from the assessment of the actual or potential failures of each method, to avoid the pitfalls in the technique chosen for a particular fabrication requirement. Similarly, even though a particular joining method might not be directly applicable in a given situation, very often an understanding of its technical aspects will aid the optimization of the method which is finally chosen.

Chapter 8

CRITIQUE OF CHIP JOINING TECHNIQUES

Introduction. This chapter supplements the previous one, and continues by providing critical analyses of some of the methods described there.

Just as a survey has generic limitations and is subject to dispute, this comparison of methods also inherently has short-comings and pitfalls. Controlled experiments (which are difficult to design) are normally required to make valid and meaningful comparisons. Yet, here we find that intuitive leaps and combinations of inductive and deductive reasoning must be used, due to the lack of such controlled and organized data. Each manufacturer or experimenter typically emphasizes a single format or process; it is quite rare that statistically valid comparisons are made of several methods in order to make a choice and if such have been done the literature has not apparently been informed. This lack of direct comparison is also due to the complexity of making statistically valid comparisons on processes as complex as those being discussed here; each minor perturbation may be quite meaningful with respect to cost, reliability, practicality, etc. Furthermore, although certain specifications do exist for aspects of stress testing and reliability, there is truly no universal standard to which electronic structures must be compared. Thus, the criteria used here must be somewhat arbitrary and subjective. However, as with the survey, even with these barriers to logical purity and statistical validity it is nevertheless considered worthwhile to present such a critical analysis - if for no other reason but to excite discussion and prod the proper experiments into existence, which will prove or disprove individual aspects of contention.

It is recognized that other factors beyond technical superiority are sometimes used by manufacturers in choosing production schemes. Government contracts, for example, will sometimes insist on a particular scheme, and once the tooling has been established for a contract and the experience attained, there is a tendency for that procedure to be continued, due to inertia and cost factors. Device or component availability is often pertinent in today's market; the smaller manufacturer generally adapts his production techniques to the components he can buy. This probably has made experimentation with flip chips particularly difficult, since devices with pads have been difficult to obtain over the last several years. There are also psychological factors involved, whereby a solid technical image must be presented to customers, and admissions of non-optimum technical directions are anathema. This can have the effect of elongating the usage of obsoleted technologies.

Because of the enormous number of potentially usable chip-joining schemes, this discussion is restricted to a group of six methods which were chosen for two main reasons: they have been used in commercial production operations and are in the public eye, and they represent in a microcosm a wide gamut of processing, formats, and materials. It is believed that these six schemes provide a sound tutorial platform and permit this discussion to stand as a yardstick for other schemes not described. Where the literature is deficient, such designations as "probably", "usually", "should", etc. will be used to distinguish such conclusions from those backed by data or direct knowledge.

The six main schemes are shown in Fig. 1. They include two back-joined configurations (Wire bonding and Imbedded devices), three flip-chip variations (Controlled Collapse, Non-molten pads, and Spider bonding,[143]), and beam leads. Although it is an oversimplification, Non-molten pads represent any flip chip configuration which involves a relatively rigid terminal pad, such as thermocompression bonds or ultrasonic bonds. The flipped beam lead-gold configuration is chosen because it seems to be the leading contender at this point. Spider bonding is chosen to represent the many possibilities of lead frame uses.

The definitions and description of the criteria which will be used for assessing these six typical types of chip-joining schemes are discussed. The

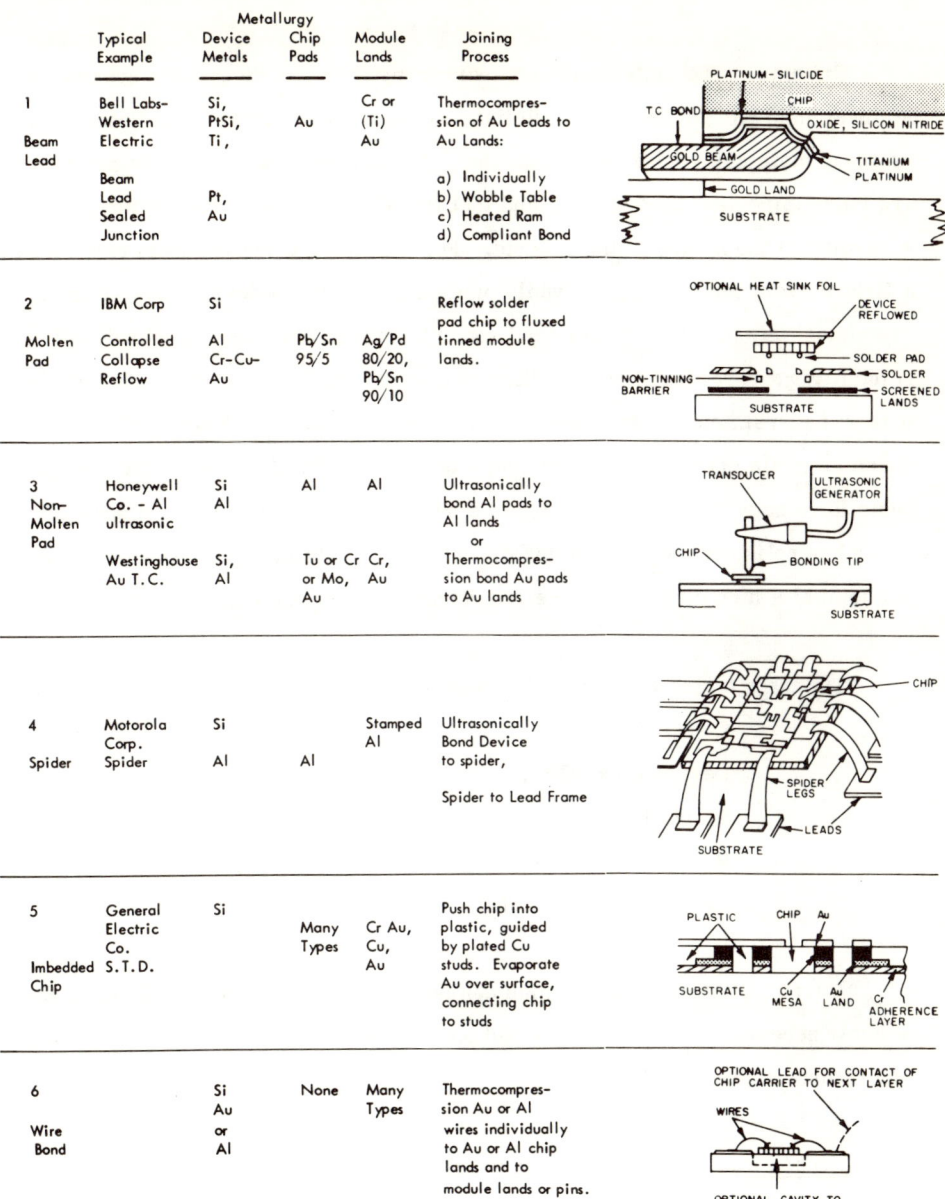

	Typical Example	Metallurgy Device Metals	Chip Pads	Module Lands	Joining Process
1 Beam Lead	Bell Labs–Western Electric Beam Lead Sealed Junction	Si, PtSi, Ti, Pt, Au	Au	Cr or (Ti) Au	Thermocompression of Au Leads to Au Lands: a) Individually b) Wobble Table c) Heated Ram d) Compliant Bond
2 Molten Pad	IBM Corp Controlled Collapse Reflow	Si Al Cr-Cu- Au	Pb/Sn 95/5	Ag/Pd 80/20, Pb/Sn 90/10	Reflow solder pad chip to fluxed tinned module lands.
3 Non-Molten Pad	Honeywell Co. - Al ultrasonic Westinghouse Au T.C.	Si Al Si, Al	Al Tu or Cr or Mo, Au	Al Cr, Au	Ultrasonically bond Al pads to Al lands or Thermocompression bond Au pads to Au lands
4 Spider	Motorola Corp. Spider	Si Al	Al	Stamped Al	Ultrasonically Bond Device to spider, Spider to Lead Frame
5 Imbedded Chip	General Electric Co. S.T.D.	Si	Many Types	Cr Au, Cu, Au	Push chip into plastic, guided by plated Cu studs. Evaporate Au over surface, connecting chip to studs
6 Wire Bond		Si Au or Al	None	Many Types	Thermocompression Au or Al wires individually to Au or Al chip lands and to module lands or pins.

Fig. 1. Description of 6 chip joining techniques.

schemes are compared, mainly by comparative charts, supplemented by discussion wherever necessary.

Criteria. The criteria for comparing methods can be grossly considered in three main categories: Cost, Reliability, and Throughput-Yield-Practicality. As these main categories are subdivided, it becomes apparent that some criteria are more important than others, and that this degree of importance may vary according to the nature of the product. For example, some products can tolerate a failure every now and again, while other highly critical devices can afford to pay almost any price for guaranteed non-interrupted service. The trade-off of reliability vs cost hinges on the ultimate cost and difficulty of making a repair. It is an understatement that there is considerable difference in designing for a space program compared to designing a television set. In order to have some basis for discussion, however, here we will assume that a system is desired which has reliability at least in the order of 0.001% failure per thousand hours per joint at a high confidence level, in normal human ambients.

The main topics for comparisons of methods include the following:
Economics,
Design Flexibility,
Reliability,
Repairability, and
Process and Controls.

A typical list of criteria which can be subdivided into those which are absolutely necessary, those which would be nice but which are really minor considerations is presented in Table 1.

Economic Criteria. Although exact costing is a complex undertaking, dependent on many elements other than direct costs, certain factors are virtually universal. For example, it is generally preferred that the cost of chip-joining should not change very much as the number of joints varies. This of course is an impossibility, since more joints often necessitate lower yields, but it is also apparent that some methods do not require changes in tooling for changes in formats, while others do.

TABLE I. Some Typical Criteria.

GUIDE RULES

- Multi-chip fabrication requires inexpensive land repair technique.
- Spring contacts are not adequate.
- Keep stresses and masses small (thin leads).
- Match thermal expansions of materials as much as possible.
- No entrapped contaminants.
- Metal resistances $> 5 \, \mu\Omega$-cm probably acceptable.
- Au:Al joints not acceptable if high temperatures are involved.

NECESSITIES

- Meets process temperature limit defined by device.
- Parts are not fragile, can be handled.
- For multilayer ceramic, replaceable chips; second or third placement has same metallurgy and reliability - surrounding devices not affected.
- Automatable, low tool cost.
- No predictable metallurgical failure mechanisms.
- At least one flexible joint to device (unless surrounded or mechanically supported).
- Minimum pad stresses.

DESIRABLE

- Simplest device alignment - avoid buildup of alignment tolerances.
- Cheapest lead type (screened > punched > etched > vacuum?).
- At least 40°C temperature difference between processes.
- Demonstrated high reliability.
- Low cost.
- Applicable to chip carriers and multichip constructions.
- Minimum connections, but not at sacrifice of reliability.
- Fewer parts - but not if few parts are extremely expensive.
- Permits backside joint to thermal conductor.

MINOR

- Permits prior in-situ device test before committment to multichip surface.
- Can permit minor design changes.
- Cost does not directly increase with number or size of leads (yield not included).
- Permits 0.002" lands.

Perhaps the main factor in integrated circuit cost is the efficiency of the usage of the semiconductor surface. Since devices are usually processed in wafer form, the cost per wafer is fixed, and the number of good devices which can be attained from that wafer dominate the cost. It is astounding for normal sized wafers in the order of 1-3 inches in diameter how drastically a few mils change in the chip size affects the total number of good chips that can be obtained per wafer.

Yield is obviously an important consideration, but it is an extremely difficult aspect to assess without specific experience. Yield is dependent on such factors as competence, incentive, environment, and capitalization as well as the technology itself; it is sometimes difficult to attain equivalent yields from two separate facilities. Thus, yield will not be used as one of the criteria, although it is certainly important. Time spent at the chip-joining station is predictable, however. The chip-joining station is often a rate-controlling factor in the manufacturing process, and therefore, time there should be minimized. Another controllable factor is the expense of the materials used in the fabrication although the type of material is often less important than the time and effort spent in fabricating it. Thus, although gold or platinum are considerably more expensive than aluminum or silver, there can be little difference in the cost of a single module based on using either the expensive or cheaper materials. Of course, for large-scale production, the sum of such small differences makes a vast difference in the total cost of a program, but it is the fabrication cost of the material that is most pertinent. A specially etched lead frame of cheap material may be far more costly than a screened deposit of precious metal. Vacuum metallized inexpensive refractory metal may take more time and thus be more costly than metallizing with a more precious metal which evaporates faster.

Some of the other obvious economic factors include the number of actual steps and the number of parts which must be fabricated, the ability for the tooling to be utilized as fully as possible (or at least readily adaptable to other uses), the degree of equipment maintenance if the equipment must be kept sharply tuned, etc.

<u>Design Flexibility Criteria</u>. This involves the amount of freedom each method has in fitting the many requirements of varied product lines - including geometrical changes - as well as electrical functionality and adaptability to various environments. The upper size limit for chips, for example, depends on the chip-joining scheme; joints which are either flexible or ductile can take up some of the stresses induced in thermal cycling, which rigidly fastened joints cannot stand. The amount of heat any device can dissipate is dependent on the number and size of the thermal paths leading from the heat producing areas.[144] Thus, thin or long leads are not much help, while thick sturdy joints are obviously much better for thermal dissipation. Some methods are restricted either to thin film or thick film techniques, while others can readily be adapted to either; this option is sometimes needed to permit changes as technologies advance. Some methods

are more adaptable to a wide range of pad sizes and shapes and the ability to put few or many devices on a single module surface. Some of the methods are also more capable of providing pads in the center portions of the device, while others are restricted to peripheral leads. Because some joints are stronger, it is feasible to use a heat sink which adds weight and mass to the chip.

Electrical characteristics are important in the consideration of design flexibility. Device speed is dependent upon such factors as conductance, capacitance, spacings, inductance, and cross talk. Difficulty can be experienced with long parallel leads which slow down the circuit. Very often, the shortest electrical path with the highest conductivity and surrounding regions of low dielectric constant are preferred. Sometimes, therefore, the ability to make connections on very small centers becomes valuable.

There are differences in the adaptability for various uses and ambients since some of these methods permit usage at higher or lower temperatures than others. Certain packages are less susceptible to environmental factors such as humidity, radiation, etc. (quite often of course, the rest of the package is just as important in these respects.)

Reliability Criteria. These criteria are described later in considerable detail, but a few comments are pertinent here. A proper assessment or reliability should relate not only to the theoretical properties of the materials and processes and geometries, but also to the realities of reproducibility and probable contaminations. It is also an aspect on which the least amount of comparative data is available. There are so many ways of evaluating reliability, so many statistical methods for interpreting the evaluation, and so little industry-wide agreements as to which of these are actually valid and predictive of field life, that it is particularly difficult at this time to view the issue clearly. Furthermore, reliability is dependent on intended usage-such as the difference between an electronic component in a space satellite, a unit in a Viet-Nam jungle, and a computer in an air conditioned building.

Discussions of reliability generally fall into three classifications:
1. Extensive field data
2. Statistical Analyses of limited tests, and
3. Postulations based on analogy and theoretical reasonings - which may or may not have relationship to the real world.

The first type has been presented on SLT modules,[70] where more than 30 billion module hours in the field have been reported. The Controlled Collapse method reliability is extrapolated from this data by laboratory stress tests and limited field input.[67] Other extensive programs, such as Minuteman have also been reported.[145] The second type (statistical analyses of limited tests) have been made on ultrasonic and thermocompression bonds.[75] Unfortunately, the third type of discussion (postulations based on analogy and theoretical reasonings) provide the bulk of the literature.

It should be apparent-although not universally accepted-that despite metallurgical simplicity, process variability can produce reliability variations. Furthermore, the degree of control can be a prime factor, leading to the interesting possibility that managerial psychology and philosophy can outweigh technology.

Other aspects relating to reliability criteria will be discussed with the Reliability Comparison Chart.

<u>Repairability Criteria</u>. Not all constructions require repairability, but as modules become more complex-due to pressures of speed and miniaturization-this need will become greater. In any event, consideration must be made as to the economic feasibility of any repair scheme, since semiconductor parts are reaching the point where they are almost cheap enough to be thrown away. When many integrated circuits are put on a single surface however, the economic justification becomes stronger. We do not address here the repair of broken conductors on the modules, but only the techniques for removing a defective chip and replacing it with another. It is important that such replacement not disturb adjacent chips or the module itself, and that the substitute chip either has exactly the same characteristics as its neighbors, or-if the joint is slightly different

metallurgically-that the reliability aspects are sufficiently understood and reproducible. This is not an insignificant task, since multiple operations might be expected to make minor changes-either in intermetallic compound concentration or residues, due to the multiple exposures to either heat or pressure or both.

The type of repair often limits the circuit density; some schemes require merely pulling the chip from the module surface, while others require a sideways plowing motion which precludes closely placed adjacent chips. Replacement schemes also vary considerably in cost and complexity-those requiring dressing of individual pad sites may be more costly than schemes where no dressing or bulk dressing can be performed.

Process and Control Criteria. As previously noted, reliability is a function of the process and control, and quality control and assurance procedures not only sharpen up a production line, but can also weed out a defective product. Even beyond this obvious observation, it should be understood that some processes require tighter controls than others to be reproducible. Not only is it important to understand the range of tolerable process variations, but also to ascertain whether normal industrially available equipment is reasonably capable of maintaining such tolerances without excessive maintenance. The canny semiconductor manufacturer would prefer not to hire technicians with the finesse of jewelers, but rather-whenever possible-high school drop-outs. A single furnace, as an example, with careful handling might be expected to maintain a narrow temperature range; it is a far different thing to control a battery of such furnaces within the same range. Many a beautiful laboratory scheme has created hardship in a manufacturing context for this reason.

It is wise to extrapolate what will happen if the controls do go slightly askew. If the pressure in a process increases somewhat, are the chips damaged? If the temperature raises a few degrees, are sufficient brittle intermetallics formed so as to completely change the reliability ground rules? If one device is a slightly different size than the other, does this preclude automating the process?

Now that we have briefly tasted the type of criteria which can be used for making our judgments, we proceed to the actual critique of these six methods.

Discussion. This section discusses the six prototype chip joining methods by means of tables and accompanying comments. These tables relate to the major criteria just discussed, with the first one summarizing the most important characteristics. The values are comparative. Strong or weak designations imply either a comparison to the other methods, or a particularly noticeable characteristic.

Discussion of Key Characteristics. Table II presents data for general relationships and comparison of key characteristics as discussed below.

TABLE II. Key Characteristics.

	Characteristic	Beam Lead	Controlled Collapse	Non-Molten Pad	Spider	Imbedded Device	Wires
1	Chip Handling	*	S	S	*	A	S
2	Automation	A	S	A	S	A	W
3	Rapid Turnover	A	A	A	W	W	S
4	Reliability	*	S	*	*	*	A-W
5	Chip Cost	W	A	A	A	S	S
6	Joining Cost	A	S	A	A	A	W
7	Replaceability	A	A	A	W	W	W
8	Thermal Dissipation	W	A	A	A-W	A	S
9	Smallest Leads	S	W	W	W	S	A
10	Inside Connections	W	S	U	A	A	S

S - Strong
W - Weak
A - Average
* - Requires Comment
U - Unknown

Chip Handling. This pertains to the structural strength of the device and its ability to be handled relatively roughly with impunity. The Controlled Collapse and Non-molten chips, for example, can be readily probed and misaligned, and then re-oriented in a simple vibrating bowl. Similarly, devices without pads are certainly handleable, but it is considerably more difficult to align them mechanically. Generally, optical alignment is required since there are no protrusions to sense mechanically. It is commonly acceded that Beam Leads should be kept in wafer orientation and the good devices picked out by computer control in order to avoid nightmarish difficulty with mis-oriented devices, since the beams might entangle each other.[87] Beams and leads are considered more fragile than device pads.[57]

In the Spider method, once the devices are fastened to the spider, none of the leads can be drastically bent without fear of damaging the chip-to-lead joint. However, since this method must be highly automated to achieve reasonable cost, the control of the processing equipment may preclude such damage. Certainly, only production experience can illuminate this factor.

Automation. This relates both to the potential for automating the chip-joining process, and to the actual degree of demonstrated automation. For example, both the Spider and Controlled Collapse techniques have been automated to a high degree, and this degree of automation helps make these methods particularly attractive. Indeed, beam lead handling has also been automated, but at a more sophisticated level, where each chip is picked out by computer control. In this case, automation may not be a cost improvement, but perhaps even a liability. It is more difficult to automate the other processes, since the Imbedded devices and Wire-bonded devices (as previously mentioned) require optical sensing rather than mechanical alignment, and the Non-molten pads do not have the self-alignment features of the Controlled Collapse flip-chip. The Non-molten pads must be put down exactly in registration with the lands to which they are joined, while several mils of leeway are available with the Controlled Collapse method.

Since wire-bonds are commonly performed by individual operators, this is not considered to be a truly automated process, even though certain aspects of chip handling may be done by automatic machinery.

Rapid Turnover. This addresses the ease with which a few parts of each type can be made. Some methods require extensive changes in equipment and processing to adapt to different sized chips. For example, new tooling would be required to punch out the frames for the Spider if the chip size or number of pads were changed, unless the leads are made by etching techniques (which are probably more expensive). Similarly, new evaporation masks would have to be made for the imbedded devices, even if the same chip were only relocated on the surface of the module. However, wire bonding is particularly adaptable since each joint is made individually. There seems to be little major differentiation among the other methods since each requires a change in the conductor pattern, which would require new masks. However, if the Compliant Bonding technique is used with Beam Leads, the compliant member would have to be changed for each format, just as with the Spider lead frame.[146]

Reliability. Data and extrapolations show and predict good joint reliability for Controlled Collapse devices in the size range of about a hundred mils square- approximately 10^{-7}% joint failures per thousand hours due to thermal stressing.[67] Alternatively, field returns from wire bonds have been rather erratic - sometimes showing excellent results, and sometimes showing failures - probably due to the variability of operators and equipment.[147]

With the lack of reliability data from production facilities, it is difficult to provide a precise figure for beam lead potential reliability. This will probably depend on the reproduceability of grain size, which of the several available bonding methods are used (Compliant Bonding,[146] Heated Ram,[148] Wobble Table[149] individual finger bonding, welding, etc.), the amount of lift-up (or "bugging"), the degree of surface contamination, etc. However, to permit decision making, some extrapolations must be made, and it is expected that beam leads will be more reliable than individually bonded wires.[150] Due to thermal expansion problems, this author expects that Spider bonds and Imbedded devices will not show as high reliability as either beam leads or Controlled Collapse devices. Similarly, the Non-molten pads, due to their relative low ductility, are also expected to be more subject to stress cracking. Further detailed comparisons on reliability will be found in Table III.

TABLE III. Reliability Comparison.

	Characteristic	Beam	Molten	Non-Molten	Spider	Imbedded Chip	Wires
1	Bond Strength	A	S	S-A	S	*	W
2	Inspectability	A*	W*	W*	A*	W*	A*
3	Process Dependence	W-A	S	A	A	U	W
4	Minimum Connections	A	A	A	W	A	W
5	High Usable Temperature (≥ 200°C)	S	W	S	S	U	S
6	Subject to Damage During Processing	A	S	A	U-W*	A	W
7	Usable in Normal Ambients	A	A	A	A	A*	A
8	Flexible Joint	S	A	W	U	*	S
9	Simple Metallurgy	A	W	S	S	S	S
10	Probable Relative Overall Reliability	S	S	A	A	A	W

S - Strong
W - Weak
A - Average
* - Requires Comment
U - Unknown

<u>Chip Cost</u>. Imbedded and Wire bonded devices are inherently cheaper than any of the others which require special pads or joining members on the device. Beam lead devices have been described as being particularly expensive, due to the relatively long etching and plating times involved in the fabrication.[57] It should be recognized, however, that large scale production of any of these methods will probably minimize chip cost differences. If test sites are desired in the discarded kerf sections between chips, the interdigitated beams preclude their presence. In such cases, test sites might be put elsewhere on the wafer, with the loss of some potentially usable silicon; in certain cases, this can be an appreciable loss of yield.

<u>Joining Cost</u>. The following factors are particularly pertinent: yield, time at the chip-joining station, and complexity and expense of equipment. Wire bonds are particularly expensive due to the long time involved in making each joint, but yields are high since each joint is replaceable. On the other hand, the Controlled Collapse technique takes a fraction of a second for the device to be positioned. All the other methods require that the device be acted upon at the chip-joining station, creating a bottleneck which is particularly pertinent for large scale production. All thermocompression or ultrasonic bonds have a slight time delay built into the chip-joining station, generally in the order of several seconds, although this can be reduced with the proper ultrasonic bonding equipment.

<u>Replaceability</u>. This refers to the removal of a complete device and replacing it with a new one resulting in substantially the same metallurgy and structure. The Imbedded devices and wire bonded devices will be more difficult to replace than the flip chip or Beam Lead styles. Residues of gold/silicon or other eutectic back joints are generally left. With the wire bonding format, however, this may not indeed be a reliability exposure and the wires could certainly be individually replaced. The Spider devices will probably be replaceable, but since each lead will have to be lifted, it might be more complex to do this than with a flip chip which can be either remelted or sheared off. Similarly, although Beam Leads may leave a residue on the lands where the chip is removed, this residue may not interfere with subsequent chip replacement.[149]

Thermocompression or ultrasonically bonded flip chips have been removed by shearing them off, but this requires that the joint be of some intermediate strength or else the lands may be ripped from the module during the removal station.[118] The easiest replacement seems to be with the Controlled Collapse or Beam Lead systems, and the most difficult seems to be the Imbedded devices.

<u>Thermal Dissipation</u>. In general, back bonded devices are the best thermal dissipation format for intermediate power levels with air cooling. The Imbedded devices are surrounded by insulating material, and may be slightly less effective than Wire bonded devices. (These considerations are with no heat sinks on the devices.) In such a format, Spider bonded devices can be variable with

respect to thermal dissipation, depending upon whether or not they are firmly fastened to a thermally conducting substrate. A similar phenomenon is involved with Beam Leads, where if the devices are lifted above the substrate, thermal dissipation will probably be among the worst of this group. For low power densities, all of these methods are acceptable; for intermediate levels with air cooling, the back joint devices are best; for very high power levels, where immersion cooling is required, again they may all be adequate, with some advantage for the flip chip configurations.[151]

Smallest Leads. This refers to the minimum lead spacing realistically achievable by the methods. Beam Leads have been demonstrated with one mil lines, and this is the smallest spacing available today. On the other hand, the flip chip configurations are usually limited to 3 mil pads on 6 mil centers, although they can be squeezed a little beyond that. Similarly, it is difficult to fabricate lead frames with extremely small dimensions. Since Imbedded devices are metallized by either vacuum deposition or etching, they may approach the miniscule dimensions of the smallest beam leads. With wire bonding, tight spacings are hindered by operators and equipment finesse. An important consideration here is whether such extremely fine dimensions are really necessary for most formats, since extremely fine lines sometimes are not capable of carrying either the heat or current density required of many circuits.

Inside Connections. This is associated with the ability to make connections to the chip inside the periphery, (sometimes called an area array). This will tend to be more important in the future, as monolithic memories and other sophisticated MSI and LSI configurations evolve. It has been demonstrated by the Controlled Collapse technique, and certainly by wire bonding. Imbedded chips can probably have lines evaporated to make contacts to the center of the device, but sometimes these configurations require multi-level metallizing, which may be a bit difficult or expensive. It is uncertain whether Non-molten pads can make such internal connections, due to their strong need for planarity. This may be less limiting if the pads are tapered to permit the tallest to be crushed during the joining operation. Beam Leads will probably have difficulty making such internal connections, if used in their present format. Spider leads might be extended to the center of the device, but again the ability for multilevel structures to permit crossovers might be more difficult with this method than by joining a Controlled Collapse chip to a multi-level substrate.

No method is universally superior in all respects as each has its particular strong and weak points.

Discussion of Reliability Comparison Table.

Bond Strength. Bond strength, of course, varies according to geometry, so that the diameter of the contact to the chip or to the module lands and the total volume of material in the joint may dictate the ultimate strength as well as the processing and material details. However, since such geometries are defined (within limits) by the need for efficient use of the silicon, some reasonable estimations can be made. Typically, it is expected that the thermocompression and reflow flip-chip joints will be among the highest in the group, averaging somewhere between 30 and 50 grams per joint.

Since some compromise is commonly made with the strength of ultrasonic joints (which can certainly be as strong as the others) in order to shear the chip from the lands for replacement purposes without also pulling off the lands, typical strengths of 20 to 30 grams per joint are reported.[95] Spider bonds are in the order of 20-25 grams when the process is under control.[143] Wire bonds are typically the weakest of the group, showing strengths in the order of 7 to 10 grams for a typical 1 mil gold wire. However, this is often a significant variable and is one of the reasons some sections of the industry are trying to replace wire bonding. Beam leads will probably be somewhat stronger than wire thermocompression bonds for two main reasons: the flatter lead geometry, and the availability of compliant bonding which does not excessively cut into the beam. Unlike with the other five methods, it is virtually impossible to measure the bond strength of imbedded chips, since the contact leads are vacuum metalized across the chip contacts. However, thin film technology experience indicates that the initial joint can indeed be reliable and strong if the surfaces are clean. Bond strengths of the others are commonly measured destructively and with air blast techniques.

Inspectability. We have previously noted that although some joints can be optically viewed more easily than others, there is no way to verify a good joint by visual inspection. The joints which are most readily viewable include Wires, Beams, and - depending on the format - Spider bonds. Flip-chip configurations can be viewed from the side, but this is somewhat difficult. Since silicon is transparent to certain infrared wavelengths, such joints can also be viewed through the module and chip by infrared examination. The Imbedded chip joints are under the vacuum metalizing, and are uninspectable.

Process Dependence. All methods will show some dependence of reliability on variations in the process, but the molten joints appear to be among the least susceptible. Next are the Non-molten and Spider joints, since the main variable would be pressure and ultrasonic frequency, or the pressure and wide range of tolerable temperature controls required for thermocompression bonding. Since the surfaces of the Imbedded chip must be scrupulously clean, contamination could be disasterous, and the ability to consistently achieve this degree of cleanliness is unknown. Both Wire bonds and Beam Leads are potentially susceptible to excessive crushing and unclean surfaces, but they may be less critical in this respect if such techniques as compliant bonding are used which preclude the notching caused by the common wedge tool. Also, see item 11, in Table IV.

Minimum Connections. All methods except Spider bonding and Wire bonding require only one joint from the chip to the module lands. The latter two require a separate bond at the chip and at the land, thus potentially increasing the reliability exposure. This is a rather strong factor in consideration of reliability, and may be a significant detriment of these two methods.

High Usable Temperature. To avoid metallurgical reactions, molten joints are usually restricted to temperature usage considerably below the liquidus. The remaining methods seem undistinguishable in this aspect when considered metallurgically. However, there may be other limitations to high temperature usage which involve thermal expansion stresses, and these are not addressed here. It is suspected that the more rigid flip-chip joints - such as thermocompression or ultrasonic bonds - will not be usable at high temperatures for this reason, but it is unfair to downgrade them without confirmation. Similarly, little data is available on thermal cycling for the Imbedded chips. Intuitively, it seems that this would be an area of difficulty, but experiments indicate that the fluorocarbon material surrounding the chip can yield and flow slightly during thermal cycling to overcome this potential defect.

Processing Damage. Since Wire bonding is usually a manual operation, there are reports of scratching the chips, which creates real damage. Although no direct information is available, one may suspect the fragility of the Spider and the long extension of the joint down the metal leads may lead to delicacy during handling which is dependent on the processing details and equipment. Initial reports indicate that handling may not be a significant problem.[143] Beam leads can undoubtedly be probed outside the chip periphery, and if bent may be capable of being straightened. Non-molten pads are more subject to probing deformations or stresses which are unrepairable as compared to molten joints (which can reform during the reflow operation). The strongest member of this group seems to be the molten joints, which can be crushed out of shape without severe detriment.

Usable in Normal Ambients. Certainly none of these methods would be in production if they were not considered reliable under normal ambient conditions. As previously mentioned, more information is required about Imbedded chips.

Flexible Joints. Obviously, the strongest members of this group are long beam leads and wires, which are inherently flexible. It has been noted that short beams may transmit thermal expansion stresses to the chip.[152] Next come the molten joints, where solder was chosen specifically for purposes of ductility. Aluminum Non-molten joints are not considered particularly ductile, but a slight improvement might be inherent in Non-molten thermocompression gold joints. It is not known whether the thickness of the lead frames required for proper assembly will be too thick for good flexibility, and this should be investigated carefully. There is a possibility that the fluorocarbon around the

Imbedded chips can provide some flexibility to take up thermal stressing, and this should also be watched.

Simple Metallurgy. The metallurgies vary considerably in complexity; the most complex is the metallurgy of the molten joint followed by the gold Beam lead metallurgy. But, even simple aluminum-aluminum bonds are aggravating since oxide contaminations can interfere with the joints.

Probable Overall Relative Reliability. It has been noted that total reliability is dependent upon a myriad of factors which include geometries, matching of materials, and even the sophistication and size of the production operation.

Based on all of these factors, the author expects that the most reliable of the methods will be Beam leads and solder reflow-molten joints, with Wire bonds representing the lowest overall reliability - due mainly to variabilities in the production operation. With the information now at hand, it is not realistic to differentiate among the remaining three methods although initial data infers that the spiders may be slightly more reliable than Wires. Of course, this estimation is worthless if one changes the ground rules, such as requiring usage at temperature extremes or environments particularly deleterious to a specific metallurgical combination.

Discussion of Process Comparison Table.

Heat and Pressure. To establish bonds, some of the methods require heat, some pressure, and some both. If two quantities must be controlled carefully instead of only one, the potential for some variation in the process becomes greater. From this standpoint, it may be that the Imbedded chip metallizing is least critical, while the wire bonds are the most critical. Thermocompression heat control is probably less critical than with reflowed joints, and the beam lead chips may actually encounter less heat than reflowed chips if the heat is supplied by the bonding tool only to the lands.

TABLE IV. Process Comparison.

	Characteristic	Beam	Molten	Non-Molten	Spider	Imbedded Chip	Wires
1	Requires Heat Control					S - Strong	
2	Requires Pressure					W - Weak	
3	Critical Time (Sec.)					A - Average	
						* - Requires Comment	
						U - Unknown	
4	Obviously Automatable	X*	X	X	X	X - Applies	
5	Critical Planarity			X		X	
6	Critical Alignment	X		X	X	X	X
7	Useful for Small Quantities	A	A	A-W	W	W	S
8	Many Dicing Options		X	X	X		X
9	Simple Reoriented		X	X	X		
10	Relative Process Criticality	S	S	A	A	A	W
11	Joining Station is Not a Delay Point	A	S	A	A	A	W
12	Probe Pressures to Unjoined Chip Probably not Harmful	A	S	W	W	W	W
13	Relative Handleability	W	S	S	W	W	A

Critical Time. Some of the processes must be controlled within a fraction of a second, or detrimental results may occur; for example, ultrasonic bonds - if vibrated for too long a time - may become weakened. However, if the pressure and temperature in a thermocompression bond is adequately controlled, the time of pressure application is probably not too important, as would be the case with the heating of reflowed devices. In the latter case, of course, the time should be minimized to prevent intermetallic formation and possible spalling - but here the time can be controlled in parts of minutes and not of seconds.

Obviously Automatable. The only techniques which may not readily meet this requirement are the Imbedded chips and Wire bonds, and even the former - with a great deal of effort-might be made automatable. The solder reflow method has received the most automation to date, followed closely by Beam leads and Spiders. However, Beam leads seem to require a great deal of automation for any reasonable quantity to be joined economically. Computers are used to select good chips for placement,[153] and this is indeed more expensive than being able to test devices and separate them into groups with similar properties which are reoriented automatically (such as can be done with molten or non-molten devices quite readily). Mechanical handling of these devices is facilitated by the presence of bumps on the chips, but if devices are not made symmetrical, they probably can be handled without the pads. There is still some conjecture as to whether beam leads can be handled by vibrating bowls due to the intermeshing of the beams and other mechanical difficulties.[149]

Critical Planarity. It is generally accepted that thermocompression or ultrasonically bonded flip chips need excellent control of the levelness and evenness of all surfaces, often requiring expensive polished substrates.[154] Certainly some, but probably not a very wide,[95] tolerance is allowable if the pads are long and tapered and can be crushed. However, the flexible Beam leads can be extended to non-planar surfaces, and solder pads permit irregularities as great as several mils. It is not known whether Spider bonds require planar surfaces, since the leads of the frame can be individually depressed against the chip surface. Imbedded chips do not require planar surfaces if the contacts to the chips are made through etched holes in the fluorocarbon.

Critical Alignment. Certainly all chips are difficult to align because of their microscopic size, but the Controlled Collapse chips have a self-compensating factor of registration within several mils, since surface tension pulls the chip into the desired position. It is certainly easier to see the Beam leads, but this does not necessarily make alignment easier. As previously noted, infrared microscopes can be used for aligning devices with opaque bumps, even in flip chip configurations. The S. T. D. Imbedded chips are aligned against copper plated studs.[8] It is important to keep in mind that mechanical alignment equipment does not require optical viewing, and thus all non self-aligning methods become more equal.

Useful for Small Quantities. This is probably Wire bonding's strongest point, where it inherently has tremendous flexibility. On the other side of the scale, the Spider configurations and Imbedded chips are less flexible since any format change requires extensive retooling and remasking. It is somewhat dif-

175

ficult to really differentiate among the remainder, except that the Non-molten configurations may require setting up new process parameters as the chip size, pad sizes, shapes, or quantities are varied; only placement equipment settings need be changed for Controlled Collapse chips (and, of course, the masks for the conductors and chip pads).

<u>Many Dicing Options</u>. The Beam leads and Imbedded chips may require exact edge dimensional control, and thus may have to be etched. This can be expensive, and in the case of beam leads may require prior device thinning.

Flip chips are capable of using slurry saw or scribe-and-break techniques, although at least one manufacturer also etches to orient his chips from the edge.[155]

<u>Simply Reoriented</u>. This was previously mentioned, but is inserted here because it is considered important. Only chips with pads seem to be reoriented easily by vibrating bowl techniques. As previously mentioned, the need for reorienting can be bypassed by keeping the devices in orientation throughout the entire processing operation, but this is considered by many technologies to be more expensive and cumbersome.

<u>Relative Process Criticality</u>. This sums up paramters 1, 2, and 3 and is indicative of the expected uniformity of reliability when normal process variables are taken into consideration. This is related to line (3) of Table III, but the emphasis is on process and not on design and materials. It is anticipated that Beam leads with compliant bonding and the Controlled Collapse methods are the least critical with respect to process. Wire bonding was already shown to be critical in this respect, and really, there is not enough information to separate out the remainder. Some ranges for ultrasonic bonding have been published and indicate a potentially broad useful range of processing variables.[106]

<u>Joining Station is Not a Delay Point.</u> The strongest member of this group is the Controlled Collapse method where chips are positioned rapidly onto the flux covered lands. The remaining processes all require a joining operation at the station creating some delay; Wire bonds are particularly notorious for long delays.

Pressures to Unjoined Chip Probably not Harmful. The strongest members of this group are the molten pads since they can reform and heal any compressed pad regions after electrical testing and probing. There may be some scratching or slight crushing of gold beams, but this will not tend to be a serious problem in most circumstances. The gold beams can be probed without any contact with the chip itself, whereas with the other methods, some damage can occur due to probe slipping and scratching of the metallization. Of course, glass passivated chips are less susceptible to such damage. On occasion, ultrasonic or thermocompression tools on the chip have been observed to cause damage.

Relative Handleability. Generally, flip chips can be relatively roughly handled or mixed up, while this will probably not be true with Beams leads, as previously mentioned. The strongest members of this group seem to be the flip chip configurations and pad-less devices.

Discussion of Design Flexibility Comparison Chart (Table V).

Power Dissipation - Air Cooled. It is difficult to precisely define low, medium, or high power characteristics, since the heat sinking characteristics, module arrangements, and manner in which air circulates past the system have as much pertinence as the power input per volume. However, it is important that heat be removed from the sections where it is being created as rapidly as possible to avoid device damage. Isothermal chip surfaces are the most desirable.

None of these systems are completely acceptable when air cooled, with no heatsinks at high power densities-in the order of over 1 watt or so per chip; conversely, all are acceptable for low power dissipation. At the intermediate power levels some differentiation can be seen, and in this case the back joined configurations tend to be superior because heat is pulled directly from a large area of the chip. Next in order of acceptability are the devices with large pads, such as the molten or Non molten joints. The weakest of this group possess thin leads which provide a high thermal resistance-Beam leads and Spiders. As previously mentioned, if heat sinks or filled plastics are placed against the chip surfaces, these ground rules are no longer valid-but this involves extra cost.

Furthermore, the designer must be extremely careful of excessive stresses when devices are surrounded with materials that expand at substantially different rates than the substrate or chip.

TABLE V. Design Flexibility.

	Characteristic		Beam	Molten	Non-Molten	Spider	Imbedded Chip	Wires
1	Power	Low	A	A	A	A	A	A
	Dissipation	Med.	W	A	A	W	S	S
	Air Cooled	High.	W	W	W	W	W	W
2	High Power, Liquid Cooled		A	A	A	A	A	A
3	Perimeter Pads		A	A	A	A	A	A
4	Internal Pads		W	S	A	W	A	S
5	Can Permit Direct Heat Sink		W*	A	A	A*	W	W
6	Short Electrical Path		A	A	A	W	A	W
7	Back Joint or Flip-Chip		A	W	W	W	W	W
8	Thin or Thick Films		A	A	W	A	W	A
9	Narrow I.O. Spacings		S	A	A	A	S	A
10	Requires Thin Wafer		W	A	A	A	A	A
11	Amount of Semiconductor Real Estate		A	A	A	A	A	A
12	Air Isolation		S	W	W	W	W	W
13	Many Formats for a Module		A	S	A	W	W	S
14	Possible Inductance Problems		A*	A	A	W	A	W

S - Strong
W - Weak
A - Average
* - Requires Comment

<u>High Power, Liquid Cooled</u>. For special circumstances requiring the modules to be immersed in a coolant, it will probably be found that there is little differentiation among any of these methods. Boiling fluid will be most effective for pulling heat from the device when the total contact area is maximized. Thus, some advantages might be inherent in flip chip configuration, but the results are not all available.

Perimeter Pads. Obviously, any of these techniques are capable of interconnecting pads around the periphery of the chip.

Internal Pads. With some methods, it is considerably more difficult to connect pads not directly accessible from the edge. For example, although an argument can be made that air isolation techniques permit such internal contacts with Beam leads, the author believes this to be an excessively difficult technique. Similarly, although some of the leads of the Spider can be extended into the device center, it may be difficult to achieve certain configurations because of the inherent need to criss-cross some of the leads which could set up problems with electrical shorting. As previously mentioned, the main problem area with Non molten pads is the planability requirement, and it was noted that an ultrasonically bonded chip may swivel under the anvil if some of the pads are too high. With proper controls and greater expense, the planarity might be controlled enough to permit this to be feasible. Imbedded chips are capable of having multiple layers metalized over them to establish electrical contact, but this may provide a distinct yield loss at the worst possible time-when the entire module is committed with chips. The strongest techniques for internal pad capability appears to be the Molten-pads-which are lenient with respect to planarity, and Wires-which can be bonded anywhere. In the latter case, some configurations require great care so the wires do not touch.

Direct Heat Sink Attachment. It is difficult to bond a metallic heat sink to a chip which has already been bonded on its back side, such as the Imbedded chips or Wire bonded devices. Of course, a case could be made that such devices are already joined to a heat sink, but it is realistic to note that ceramic materials other than beryllia are at least an order of magnitude less thermally conductive than metallic heat sinks. Undoubtedly, heat sinks can be attached to Beam leaded devices, but there is conjecture whether the thin beams are capable of supporting the extra mass-particularly when shock or vibration tests are applied. The remainder of the group are acceptable in this respect, and heat sinks more than 30 or 40 times heavier than the chip can be applied to the Controlled Collapse devices without appreciable change in the chip-to-substrate height.[156]

Short Electrical Path. Since speed is a function of path length, the strongest members of this group have the shortest paths and it is difficult to really separate Beam leads, Molten joints, Non-molten joints, or Imbedded chips in this respect. However, it would seem that both the Spider leads and Wires tend to be substantially longer than the others. Of course, this may not be the case, depending on the electrical path from the joints out to the next connection, and this second conductor pattern may be just as critical. Presently, this is not considered to be a particularly important aspect.

Back Joint or Flip Chip. Only the beam leads seem to be readily adaptable to any configuration, but the leads may have to be elongated appreciably if joined back down on a flat substrate, causing a possible severe sacrifice of silicon real estate. The remainder are considerably more constrained, and require more engineering to accomplish such flexibility. For example, solder pad devices can be joined back down and interconnected by means of decals, overlays, or straps- but some of the simplicity of the original reflow process has been sacrificed. It is difficult to conceive how Wire bonded chips can be joined face down.

Thin or Thick Films. Only two techniques-the Non-molten pads (again due to planarity constraints) and the Imbedded chips (requiring vacuum metallizing for interconnections)-are restricted to only thin films; the remainder have considerable flexibility in this respect. The Spiders, for example, can be attached to thick or thin films on the modules, as can Wires.

Narrow I/O Spacings. The Beam leads and Imbedded chips are probably capable of the smallest spacings of the group, but this differentiation is in the order of several mils. Although for certain special circuits this may be pertinent, the author does not believe that it is important in the majority of applications.

Requires Thin Wafer. Any of the processes requiring etching for separation from the wafer will normally require a thin wafer-such as beam leads and the Hughes flip-chip. The remainder of the methods are more flexible. Thin wafers may be advantageous for heat dissipation but tend to warp with multilevel metallization; thus they can be advantageous or not, depending on the format. In any case, they may be somewhat more costly due to the extra processing and handling problems.

<u>Amount of Semiconductor Real Estate</u>. Contrary to some opinions, the author believes that there is no great difference among any of these methods. Before conclusions can be drawn with respect to pad contact area, amount of kerf required for dicing, etc., the specific details of the remainder of the device must be known. Thus, in individual cases, one method might be slightly advantageous over the others, but it is difficult to predict. However, the interdigitated Beam leads do not readily permit the location of test sites or markings in the kerfs; these can be useful and their sacrifice is unfortunate, particularly as devices get more complex. If test sites are essential, Beam lead configurations may use up valuable space to contain them. Very narrow Beam spacings cannot be interdigitated and thus consume valuable silicon real estate.

<u>Air Isolation</u>. Only the Beam Leads are capable of air isolation, but although this was described as a significant advantage, the industry does not seem to have responded particularly, perhaps because of the large silicon real estate loss.

<u>Many Formats for a Module</u>. The strongest technique for this type of flexibility appears to be the Molten joints; there is no change in the process for chips of different size or shape or intermixes-all that needs to be done is for the chip replacement machine (or operator) to position each chip in the proper place. Of course, the land pattern on the module must be changed-but this is a universal requirement-except perhaps for Wire bonding, which is the next strongest of this group. The beam leads and non-molten joints may require some change in the bonding parameters to adjust for different numbers of pads or chip sizes, with the exception of the compliant bonding-which requires a new throw-away compliant member. It would seem that the Spider and Imbedded chip methods are the weakest of this group; the former requires a new lead frame, while the latter requires different placement of the studs on the substrate as well as the vacuum metallized interconnection pattern.

Possible Inductance Problems. There are two edges to this sword: some electrical designers worry about the inductance potentially inherent in an extended lead, while others enjoy the ability to be able to doctor the inductance by changing the lead length. Thus, Beam Leads are either a joy or a nightmare depending on the point of view and requirements. There does not seem to be a major problem with bulky pads, but the Spiders and Wires do run the risk of long lead length which could cause a problem.

In summation of Table V, it is evident that each method has its own advantages and disadvantages with respect to flexibility, and it is erroneous to approach a technology merely because it is extendable in a particular direction (unless, of course, it can be established that it is the exact direction required by a particular manufacturer). With a constantly changing integrated circuit technology, it requires considerable astuteness to predict the path of flexibility which would be most pertinent to future generations of devices.

Chapter 9

POWDER INTERCONNECTIONS

Introduction. As microelectronic modules become more sophisticated, the ability to interconnect different layers of metallizing becomes more critical. Several methods have been used rather routinely for establishing such interconnections: printed circuits on organic substrates typically utilize plated through-holes (or vias); multilevel ceramics are constructed by screening conductive pastes into pre-punched holes,[7] yellow wire and interstitial pins are also means for interconnecting metallizing layers.

A somewhat different method is described here which has particular pertinence to multilevel ceramic modules and to interconnecting two sides of a ceramic substrate. The process simply consists of filling the open cavities with conductive powder and fusing the powder to the module metallizing to form reliable interconnections. The advantages, limitations, materials, techniques, variations, and properties of this process are briefly discussed. As before, the discussion is restricted to thick film ceramic modules either of the SLT type[61] or multilevel type.

The interconnections are formed in the following manner: via holes which are closed at the bottom are filled with flowable metallic powder which electrically connects to internal conductive layers or other electrical parts. The module is then fired to sinter the powder and metallurgically bond it to the electrical parts which it contacts. Top lands can then be screened on the ceramic surface contacting the fused powder, and fired. If the fired lands are then dip tinned, the molten solder can permeate the powder and form a solid metal plug.

Two types of internal land connections are thus attainable: a butt type, where the powder contacts the lands at the bottom of the holes (Fig. 1), and an edge type, where the contacting land extends into the side of the hole but allows the powder to fall past to contact other lands (Fig. 2). Obviously, several edge connections can be made in a single hole, but only one butt connection (unless, of course, the butt lands are segmented). In addition, several configurations are attainable, such as: transverse internal channels can also be simultaneously filled with the vertical holes, resulting in three sections of a winding (Fig. 3); two sides of an SLT type module can be electrically joined (Fig. 4); very deep holes can be reliably filled, providing very high length-to-width ratios (Fig. 5). These various types and configurations will be further described shortly.

Several methods of powder filling are: a) pressing the module surface into the powder; b) vibrating the powder on the module surface into the holes; c) brushing the powder into the holes, d) cascading the powder into the holes. Method b) (vibrating) has been found to be particularly easy, and was used for the majority of the experiments in this chapter. However method d) (cascading) is probably a more automatable process. With the proper size powder, the filling method does not seem to be critical; all of these methods are satisfactory, and only large scale evaluations can differentiate yield, economics, or fabrication ease parameters.

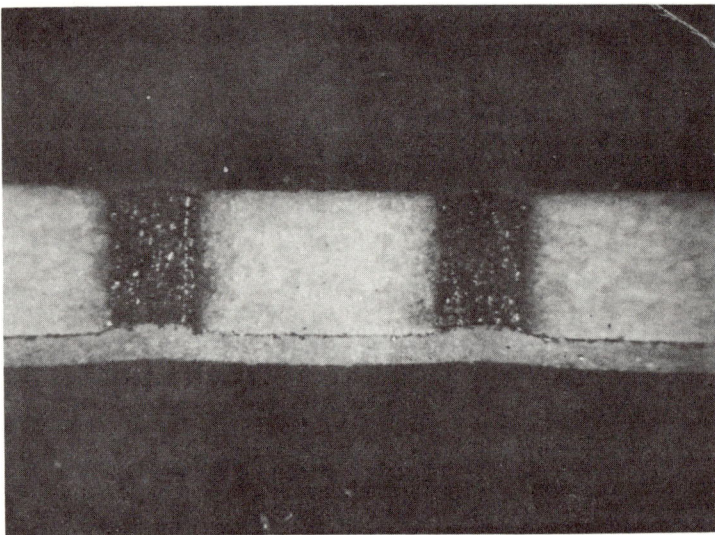

Fig. 1. Butt connection to internal silver land with soldered silver power.

Fig. 2. Edge connection to internal silver land with soldered silver power.

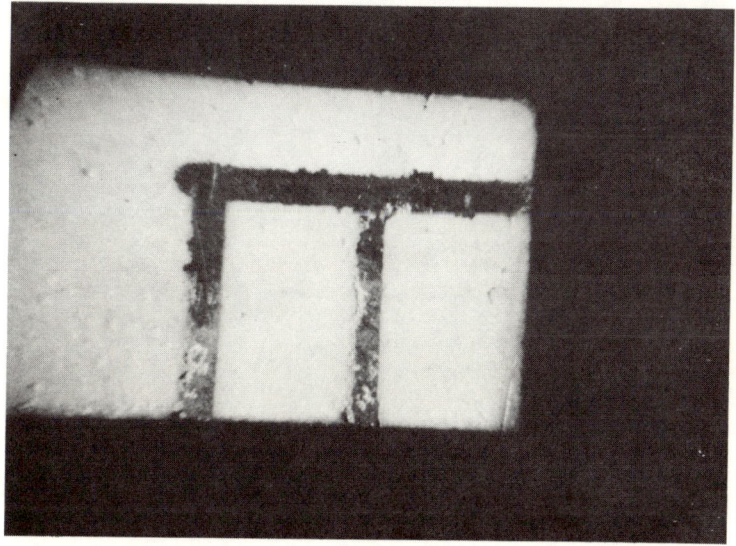

Fig. 3. Transverse filled section, soldered (each section approximately 1/4" long).

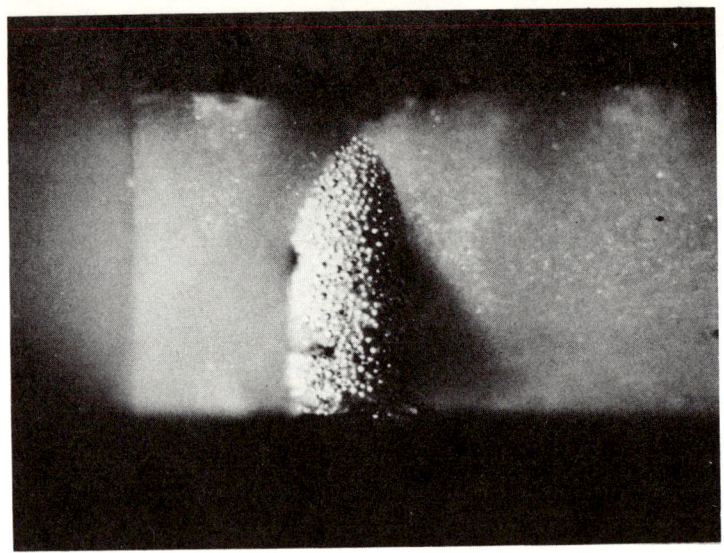

Fig. 4. Unsoldered silver spheres joining both sides of SLT substrate. (Module cracked to show filled-in hole.)

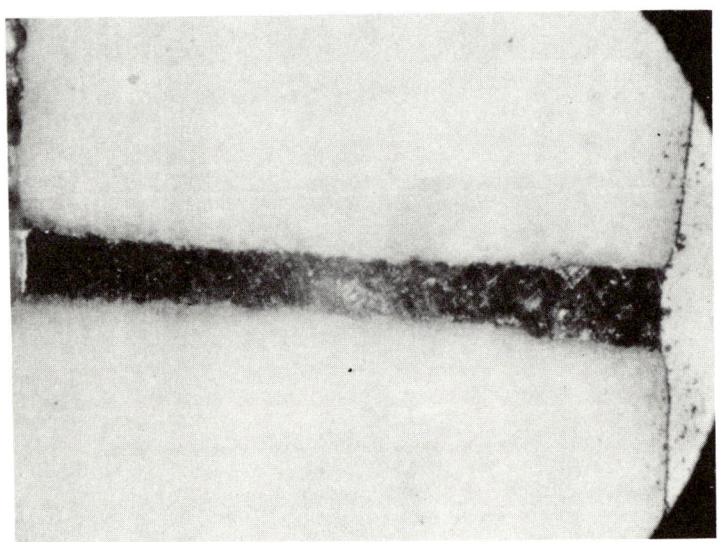

Fig. 5. 100 mil long via filled with soldered silver spheres contacting internal silver land.

Advantages and Disadvantages. The chief alternative method for electrically interconnecting the levels in multilayer ceramics involves screening each individual green (unfired) ceramic slip layer at least twice -- once for applying the land format and partially filling each hole with paste, and again to completely fill the holes from the other side. With slip thicknesses of 5 mils and greater, it does not seem that this process can be combined into one step with present equipment if high yield reliable parts are desired. Some preliminary experiments attained low yields when one pass was used, combined with subsequent capillary soaking with liquid copper[157] but this might be improved. With powder filling, only one pass is needed -- for the lands -- and with modules requiring many layers, this is a substantial economy. Filling holes with paste requires accurate registration; powder filling requires none. In fact, no change in tooling is required as circuit formats are changed other than the land pattern masks and hole punches.

Powder filling has shown exceptionally high yields -- so good that no yield estimations can yet be made; the process is very close to 100% with holes 10 mils in diameter or greater (based on thousands of holes). However, if powder should not enter a hole, the part can be re-processed without any detrimental effects. In fact, even if an internal clog in a hole should create an electrical open, the powder can be pressed down in that hole and reprocessed to salvage the part. It is doubtful that such repair or regeneration can be applied to paste filled vias.

Another important advantage is the minimizing of the number of good contacts. With paste filling, each layer must be perfect -- multiplying the number of required good hole fillings by the number of layers. Even with a high yield process, large numbers of layers could preclude high yield finished modules. Consider for example, a module 20 layers thick with about 100 vias. With an assumed arbitrary paste hole filling yield of 99.9% it is still likely that one of those holes will be defective. (Assume 2,000 holes, although in actuality it would be less since some holes probably would terminate in the upper portions of the module.) In a similar construction with powder filling, only about 100 internal contacts need be established.

A few limitations do apply to powder filling: 1) all via holes must be open at the top surface; no blind holes should be made; 2) for adequate filling, holes should be about 0.008 - 0.010" minimum diameter; 3) the holes should be closed at the bottom to prevent the powder from falling through.

The importance of these limitations should be assessed by balancing module requirements against fabrication cost advantages of the method. It is conceivable that a few layers be paste filled, for example, to achieve blind vias, and the remainder powder-filled to minimize costs. The powder filling might also be combined with a subsequent capillary soak technique; dip tinning is such a process (Fig. 6).

The advantages and disadvantages of this method as compared to two-pass paste filling are listed in Table 1.

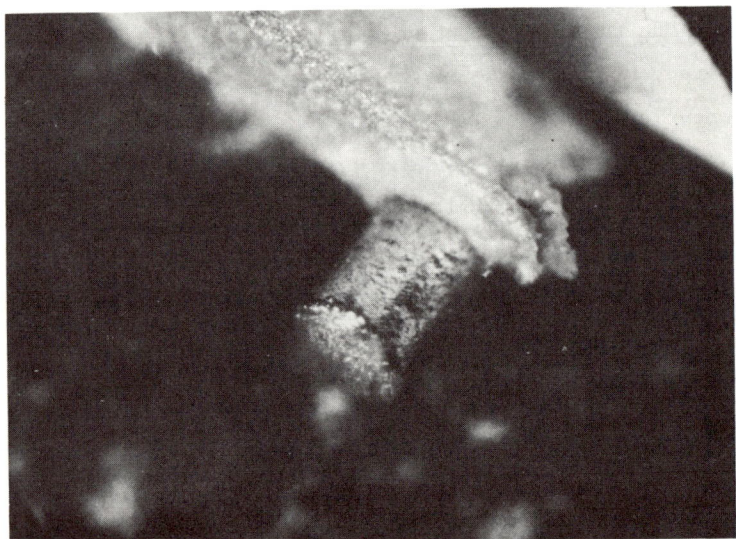

Fig. 6. Solder - permeated plug still adhering to inside land after module was broken apart.

Materials. Two types of conductive materials require consideration: screened lands and powders. Both of these materials of course, must be compatible with the ceramic materials. For example, multilevel ceramic constructions which can be fired at low temperatures (under $950^{\circ}C$) permit the use of

Table 1. Advantages and Disadvantages.

Advantages	Disadvantages
• Low cost	• Limited to about 0.010" holes minimum
• No registration	• No blind holes
• Easy format changes	• Possibly more susceptible to thermal shock
• Different diameter holes in same module	• Holes must be closed at bottoms
• Cofiring or postfiring	
• Excellent depth-width ratio	
• Can produce transverse sections	
• Excellent yield	
• Yield improved further by limited contacts	
• Repairable	
• Permits replacement of interstitial pins at low cost	

gold, silver, and silver-palladium internal lands (and, potentially, copper), while high temperature ceramics (such as alumina) permit palladium, molybdenum, tungsten, etc. The conductive powders must then be matched to these lands and ceramics. Thus, silver, gold, silver-palladium, silver-magnesium, and gold-platinum powders were used for connections in low temperature ceramics, and pre-alloyed silver-palladium spheres were successful in joining to internal palladium lands in higher temperature alumina ceramics (Fig. 7). The powders used in the experiments and their compatibility with both internal lands or top lands is summarized in Table 2. The rather extensive possibilities become apparent here, as well as the need for proper processing conditions.

Small particle powders (about 15 microns or less) shrink drastically in volume when fired in the temperature range demanded ($\geq 650^\circ C$) by the top lands Such small particles also tend to bridge, stick and give difficulty in filling nar-

Fig. 7. Cross - section of soldered Ag:Pd spheres contacting internal Pd land in alumina ceramic.

row holes. Thus, particles from about 1-3 mils fill very well; a range of 15-60 microns could be considered optimum. Furthermore, certain powder types are preferred for specific methods. For example, non-free-flowing powders pack better when the module is pressed into them. However, large non-sticking spheres (as gold) fall out of the holes if they are turned upside down, and therefore are best used with vibration or cascading. Multilevel alumina ceramics require high melting internal metals, since they must generally be fired above 1300°C. Palladium is one example, providing reasonable conductivity (typically $0.5 - 0.9 \Omega$/inch/10 mil line) without requiring the reducing atmospheres needed for Mo or W. However, initial attempts to establish electrical contacts with silver powder to internal palladium lands were unsuccessful: the poor contact permitted many electrical opens, particularly after the top lands were fired. Cross-sections of such joints indicated a slight withdrawal of the sintered plug away from the palladium land as the module cooled from the sintering or top land firing temperature. This was true of Ag powder with sintering temperatures of 700°C, 750°C, 800°C and 850°C -- all for twenty minutes. Also, a mixture of 80% Ag spheres and 20% Pd spheres did not establish good contact.

Table 2. Summation of Powders and Processes.

Powder	Type	Processing	Au	Contact to Internal Lands Ag, Ag:Pd	Pd	Contact to Top Lands	
Au	H & H fine precipitate 200-100 M spherical HA brown	G 750°-10, E SH 825°-5 SH	G	F	G	G	Ag:Pd
Au + Ag	Mixtures of Au and Ag spheres	G 750°-10		G	G	G	Ag:Pd
Ag	H & H Silpowder 120 400-200 M Spherical 400-200	G 700°-10 SH 825°-10 G 700°-10	P P	G G	P	P	Au
Ag:Au	H & H BT eutectic (28 Cu:72 Ag) 200-400	G 700°-10 Tinning P G 700° in N$_2$		G			
Ag:Pd	H & H Premalloy 800 (80 Ag:20 Pd) 400-200	G 1000°-5		G	G	G G G	Ag:Pd[3] Au:Pt:Pd[4] Au:Pt
Ag on Pt	M 25% Ag on Pt	G P25°-10		G			
Ag on Cu	H & H 5% Silpowder 200-100 M 3% Ag on Cu	G 100°-10 oxidation in air Tinning P		F			
Au on Cu	In-house deposition	G 700°-10 tinning P	G		G		
Au:Pt	In-house deposition 80 Au:20 Pt	G 825°-10					
Ag:Mg	SP 1% Mg:Ag SP 9% Mg:Ag	G 700°-10 Tinning P 700°-10		G		G	Ag:Pd
Ni	Mackay Ni	P 700°-10 S Tinning P		P			

KEY:		Powder Sources:	700° sintering temp.
G	Good	H & H - Handy & Harman	-10 minutes at temp.
P	Poor	SP - Spex	200-400-particles between 200
F	Fair	M - Metz	and 400 mesh
E	Erodes	HA - Hanovia	
S	Sintering		
SH	Shrinkage		

However, Au, mixed spheres of 2 Au: 1 Ag, and pre-alloyed 80 Ag: 20 Pd (Premalloy 800 flame sprayed spheres from Handy and Harman Co.) establish good electrical contact to internal palladium lands, Fig. 6. As expected, the Au and Au:Ag erode quickly in solder, while the Ag:Pd dissolves much more slowly. These systems seem relatively insensitive to the top lands: good top contacts were established with Au:Pt, Au:Pt:Pd and Ag:Pd lands with no opens. However, the ternary electrode contracted more when fired on the gold and Ag:Pd powders, pulling the filling toward the land. This might create undesirable stresses and may suggest balancing top lands around the vias to avoid such pulling effects with any top land.

The fusing temperature of the powder is dependent on the sintering tendency and melting point of the metal or alloy used -- increasing with increasing melting point, but limited by the ceramic requirements. Some powders also provide good contacts if fired in inert atmospheres (such as nitrogen) to avoid oxidation: BT silver-copper eutectic, 3% and 5% silver-plated copper, and gold-plated copper have provided good contacts. However, common tinnable top land formulations do not adhere well to ceramic surfaces when fired in non-oxidizing atmospheres. Thus additional development is required if such a system is desired to be fired in a nitrogen atmosphere.

Processing. Two alternative methods are known for applying the top lands in multilevel ceramic modules: screening the green sheet and cofiring it with the ceramic, or firing the ceramic first and then screening and separately firing the top land. Each has its disadvantages. Cofiring requires a difficult matching of the metallurgy and ceramic and slight differential shrinkage differences make subsequent alignment process (as chip joining, for example) potentially more difficult. Post-firing overcomes these disadvantages, but requires screening on what may not be as flat or smooth a surface, and may jeopardize the entire expensive part if a screening defect occurs. Fortunately, powder filling can be used with either technique. For post-fired top lands, the powder can either be sintered with the ceramic firing or alternately fired afterward, and then the lands applied and fired. For co-fired top lands, a slight bit of trickery is required in that the top lands are not initially screened over the holes but only in the circuit pattern. The powder again can be co-or-post-fired, and then contacted with more land material. Covering the sintered powder with spots of electrode paste like this is an extremely high yield process, capable of ready touch-up if required -- unlike the screening of fine lines.

It might seem that cleaning the extraneous powder from the surfaces of the module should be difficult, but this is not so if spheres of the desired size are used. Simple wiping procedures before sintering clean the surfaces quite well without pulling the powder out of the holes. When cascading powder is used together with vacuum-cleaned counter-rotating brushes, this can be accomplished automatically. Dip soldering also dissolves extraneous gold or silver. After sintering, controlled brushing procedures can also be utilized for cleaning off extraneous particles, but if any extraneous particles are co-fired with the ceramic, or pressed into it, they might be imbedded firmly and difficult to remove. No need has been indicated for either incorporating binders into the powder, or compacting the powder into the holes, but these are logical extensions if desired. Vibrating tables are adequate for settling the powders densely and compactly into the cavities.

A further application of this method is filling closed bottom cavities with solder. If solder powder is poured in, melted, and then cooled, solid solder can be made to join to the lands at the bottom (or edges) of the hole. Care must be taken either to prevent oxide formation on the solder powder, or to perform the operation in a reducing atmosphere; otherwise the particles will not coalesce. Another interesting possible application for this technique is the fabrication of chip capacitors. To avoid the difficulty of establishing end contacts in these multilevel constructions, internal vias can be used to contact the internal planes.[158] Powder filling these vias might yield economical constructions which are more sturdy than edge metallized units. Of course, format, dielectric real estate, and equipment conversion costs must be carefully considered for such a usage. Fabrication of transformers may also be realistic when using powder filled vias for most of the windings, connected by a screened top layer.

<u>Reliability Under Severe Stress</u>. A series of evaluations with several different types of thermal and mechanical stresses were applied to limited numbers of samples with a diversity of formats and materials. Resistance changes on some conditions were observed, but the reliability exposure (if any) was not rigorously assessed. The following is typical of the results:

a. 240 via holes with tinned lands were put through a chip joining furnace at 340°C ten times with no discernible visual changes or electrical opens.

b. After 84 cycles from -60°C to $+125^\circ$C, minimal changes were noted with contacts to internal silver ground planes: Ag powder increased 0.8-4 milliohms from initial readings of 35-75 milliohms; gold powder increased 0.6-7.0 milliohms from about the same range; Ag:Mg powder increased 0.4-19.9 milliohms. Gold ground plane samples were much more erratic with the same powders.

c. After sinusoidal vibration at 50 G's, constant peak acceleration between 46 and 20,000 Hz (20 minute sweep) perpendicular to the substrate, no opens were observed with continuous monitoring. Of 8 Ag filled holes and 8 Au filled holes (contacting internal Ag and top Ag:Pd lands), only one pair of Au filled holes showed a 10 milliohm increase. The same samples were submitted to impact shock at 1,000, 3,000 and 45,000 Gs. No changes in the Au filled holes were observed, and less than a 5% resistance change was noted in the silver filled holes. These specimens seemed to show good mechanical stability under these conditions.

d. After 100 cycles from -60°C to $+150^{\circ}$C, one Ag edge specimen (40 holes) showed resistance changes less than 10 milliohms. 20% of the butt connection holes showed between 10 and 250 milliohm increases -- only 2 of 40 holes showed a 100 milliohm increase. After 500 cycles however, the butt connection module had 3 opens and 1 intermittent open. At 750 cycles, the edge specimen also had 3 opens. All but two of the six opens were from cracks in the top lands, particularly where they passed the hole edge and were weak. Sizeable resistance increases up to several ohms were measured in both types after 500 cycles. Another butt specimen stored at 150°C for 1000 hours showed negligible changes.

Very severe thermal cycling does create trouble with such constructions, but not right away, and such cycles may create failures which could not occur under normal usage conditions. In this instance, the structure of the top lands was degraded -- which is not normally expected. This aspect obviously requires further study.

e. Premalloy 800 Ag:Pd contacts to internal Pd lands were crudely stressed as follows:
   1. Shock: One sample was dropped from a height of 18" onto a chemical bench top one hundred times with no resistance change.

2. Solder redip: One sample was redipped for thirty seconds, with no opens occurring. Resistance increases of considerable magnitude were observed on some sections; others decreased. The changes were not diagnosed, but land erosion and changes in solder thickness are suspected.

3. Multiple reflows: Two samples were heated at 340°C for several minutes five times, with no opens. The top lands appeared very uneven -- as if the solder had disappeared in some sections. The resistance changes were quite strange; some rows showed very little change, while others showed resistance increases in the order of 20 mΩ. The main changes are attributable to interactions of the solder with the top lands. Measurements of 2 vias and the internal lands only (no top lands), show no significant changes.

4. Thermal Cycling: Modules with the Permalloy 800 Ag:Pd sintered at 750°C, 825°C, 900°C, 950°C, and 1000°C were cycled from 0°C to 125°C. After about 500 cycles, only the 1000°C samples showed no change in resistance -- the others showed changed which became more severe with decreasing sintering temperatures. When an additional 100 cycles from -40°C to $+150^{\circ}$C were run, even the 1000°C (and a module at 1050°C) showed resistance increase in some sites. It appears that modules sintered at 1000°C are acceptable for more normal temperature fluctuations, but not extremes from cold to hot.

5. 150°C Storage: No change in resistance was observed after 500 hours in a sample sintered at 1000°C. However, an additional 500 hours at 250°C produced drastic increases and even some opens. Obvious interaction with the solder and top lands were observed; this is beyond the metallurgical compatibility temperature range.

In summation, such results imply that metallurgically sound systems (such as silver:palladium pre-alloyed spheres with palladium lands) are indeed capable of withstanding rather severe stresses. Any module fabrication scheme, if stressed sufficiently, will degrade; it would not be unexpected for similar changes to occur with other fabrication schemes. Thus there does not seem to be a major reliability problem with constructions of this type, although indeed comprehensive reliability testing would always be in order before committing any specific scheme to a production operation.

Demonstration and Formats. The following demonstrations illustrate various aspects, capabilities, and variations of this process:

A. Yield. Several thousand holes of 15 to 20 mils in diameter and about 30 mils deep were filled with silver spheres to contact internal silver metallizing; no electrical opens were found.

B. Deep Holes. Sixty 15 mil diameter holes about 100 mils deep were filled with silver powder, another 60 with gold powder, and all were electrically sound. In another sampling, holes of 9 mils, 14 mils, and 40 mils diameter were drilled 280 mils deep; 4 each of the 14 mil and 40 mil diameter holes were electrically good with both silver and gold powder, while only 3 out of the 4 of the 9 mil diameter holes were good. This indicates that deep holes can indeed be filled if they are sufficiently wide.

C. Solder Penetration. These deep samples were dip tinned in 19 tin:90 lead solder at $625°F$ for 5 and 10 seconds for the 100 mils specimens and 20 seconds for the 280 mils deep specimens. The silver-filled samples were completely permeated with solder even through visually intact silver:palladium top lands. The gold samples permeated only to about 25 mils, but this is not a desirable metallurgical system anyway.

D. Very Narrow Holes. Attempts were made to fill 3-4 mil diameter holes, with very little success. Although spherical gold and silver powders enter the holes, the particles were too large in relation to the hole diameter to be of much value -- sometimes only 2 across the hole diameter. Finer powders bridged and did not fill. If it is important to use this method for filling such small holes, techniques must be developed for preventing aggregation and bridging of small particles.

E. Co-Firing With Low Temperature Ceramics. 48 holes each of silver and magnesium: silver-filled holes were electrically good when co-fired with silver internal metallizing in a low temperature ceramic formulation 150 at 825°C for about a 7-hour complete cycle. Such a co-fired system appears feasible, but the surfaces must be cleaned of extraneous powder before firing or the particles will contaminate the module surfaces.

F. <u>Contacting Both Substrate Surfaces.</u> Silver palladium electrodes were screened over the pin openings on one side of SLT type substrates; (Fig. 8) after firing, the holes were filled with silver powder and sintered at 700°C for 10 minutes. The tops of the powder filled holes were again covered with silver:palladium screening paste, and the modules refired. (Fig. 9) With these samples, 190 of 192 holes were electrically good, even under the rather crude experimental circumstances. When dip tinned, solder completely permeated the powder filling, creating a solid metal rivet (Fig. 10). This could be a high yield, low cost, flexible interconnection method. The two electrical openings were due to lack of flow of the top conductor, which merely requires an optimization of rheology.

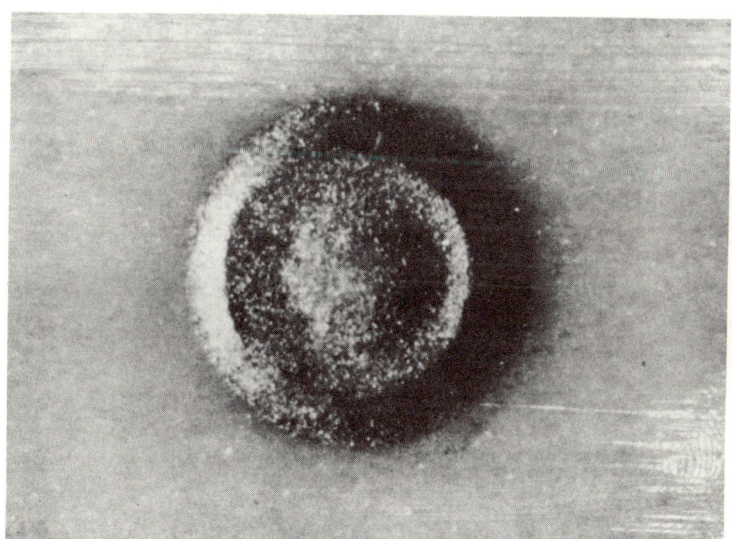

Fig. 8. Top view of screened land over SLT module pin hole.

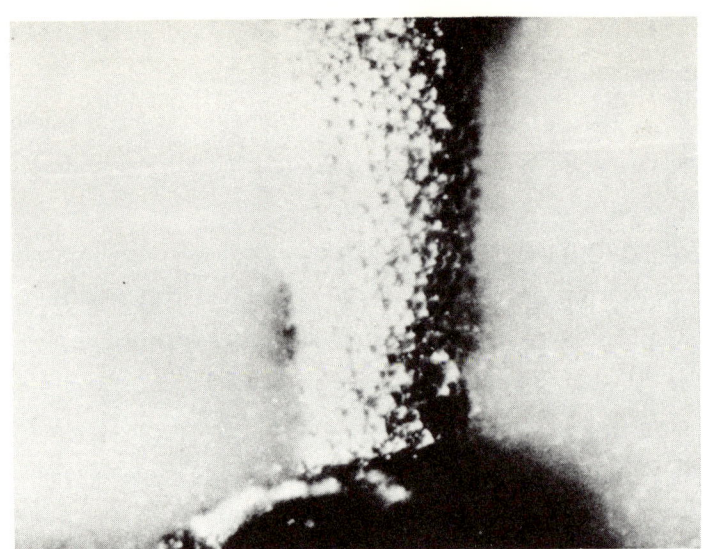

Fig. 9. Cross-section of filled-in SLT pin hole.

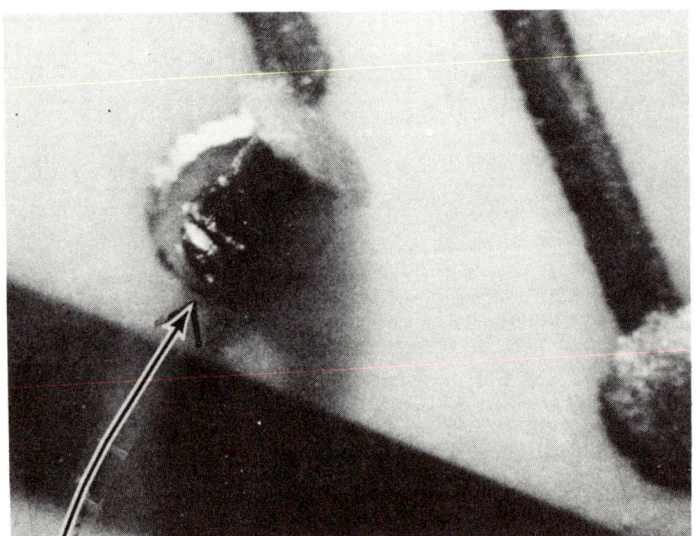

Fig. 10. Tinned plug pushed out of SLT pin hole.

G. <u>Shallow Holes</u>. Silver powder was vibrated, brushed, or pressed into holes varying from 5 to 25 mils deep, and then sintered at 700°C for 10 minutes. All of the samples were electrically good, except for a few where the multilayer ceramic layers were misaligned. No apparent problems are indicated in achieving fillings of different depth holes on the same substrate, even when several of them are shallow.

H. <u>Various Width Holes.</u> 100 percent yields were obtained with 40 holes each of nominal 6, 12, and 17 mil diameters distributed in the same parts. No problems are apparent when different size vias are used in the same part.

I. <u>Filling Traverse Channels</u>. Samples with drilled 25 mil diameter holes which included transverse sections were filled with silver spheres from the top of the holes. After sintering at 750°C, 7 U-shaped specimens, each leg about 1/4 inch long, were electrically continuous (see Fig. 3). This may be an economical process for winding transformers, cores, etc. when the top holes are joined with a final screening of the conductor material.

<u>Electrical Resistance</u>. The contact resistance of these powder fillings to internal lands approximates only a few milliohms. The electrical resistance of soldered powder vias seems to be about 4 to 10 times greater than the bulk metal, depending on the process conditions. A comparison of bulk and actual resistance values for gold and silver spheres is listed in Table 3: Mr. R. A. Delaney measured and calculated some contact resistance values for samples with spherical silver powder, internal silver lands, and various top lands. He found contact resistance varying from about 1 to 4 milliohms for all conditions, which included tinned and untinned specimens. Top lands were gold:platinum, silver:palladium, or gold:platinum:palladium. After these samples were stored at 150°C for 500 hours, the same range of resistance values was observed, with deviations of about 2 to 3 milliohms maximum, showing both increases and decreases in resistance -- a negligible amount.

Table 3. Calculated and Attained Values.

Material	Resistance of 10 mil wire (B and S gauge 30)	Resistance of 10 mil via hole and 2 contacts	Measured Apparent Density	Calculated % Metal in Hole
Spherical Ag	0.0081 ohm/in.	0.07-0.09 ohm/in.	88 gms/in^3	51%
Spherical Au	0.0127 ohm/in.	0.090-0.145 ohm/in.	175 gms/in^3	55%

Similar data was obtained with Premalloy 800 silver:palladium spheres contacting internal palladium metallizing in alumina ceramic which had been fired at 1500°C The samples were subdivided with several different types of top lands, both with and without dip tinning (Figs. 11, 12). The following trends were observed: the measured resistances covered a range of 55 to 70 milliohms, which includes a minimum of 2 squares each of top and internal land sections of screened 10 mils width, as well as 2 vias (Fig. 13, 14). The resistances decreased as the sintering temperature increased, from 53 milliohms average at 750°C to 43 milliohms average at 1000°C; the major drop occurred at 950°C. This is attributable to better particle contact at the higher temperature since the palladium land (previously fired at 1500°C with the ceramic) was unlikely to change resistance at the lower refiring temperatures. The contact resistance of the powder to the internal palladium land in every case was less than 10 milliohms.

With silver:palladium top lands, tinning generally lowered the resistance a minor amount, while with the less conductive lands (as gold:platinum, gold:platinum:palladium), tinning decreased the resistance much more significantly, as might be expected.

Estimations of the actual resistance values of each portion of the construction were obtained in several ways. Particularly thick samples were prepared with 120 mils long via holes instead of the normal 40 mils, giving a difference of about 80 mils in depth. The normal control samples averaged 56 mΩ while the longer ones averaged 107 mΩ. Assuming no difference in powder packing or sintering for the different length holes, this gives a difference of 51 mΩ for 160 mils of via length, or 0.32 mΩ/mil.

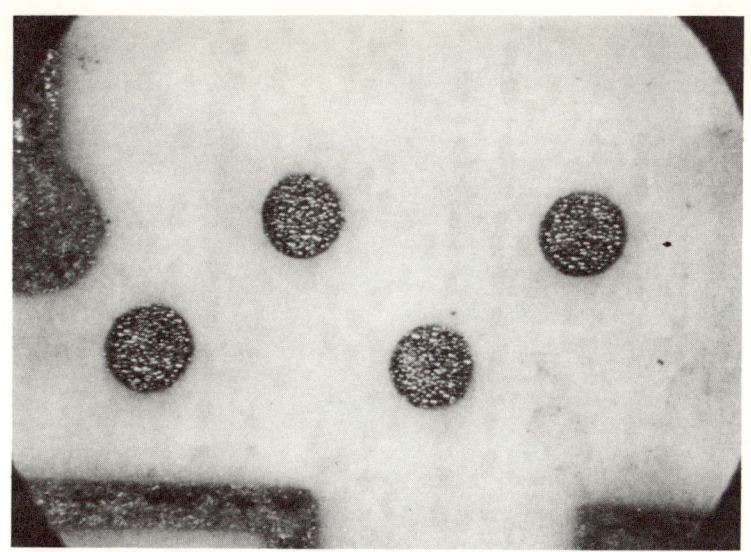

Fig. 11. Top of untinned Ag:Pd vias (no lands).

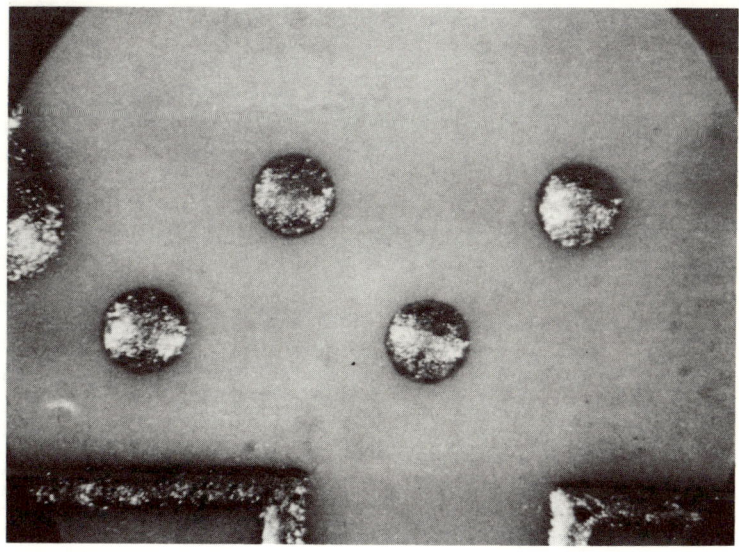

Fig. 12. Oblique x-ray view of a section of test pattern.

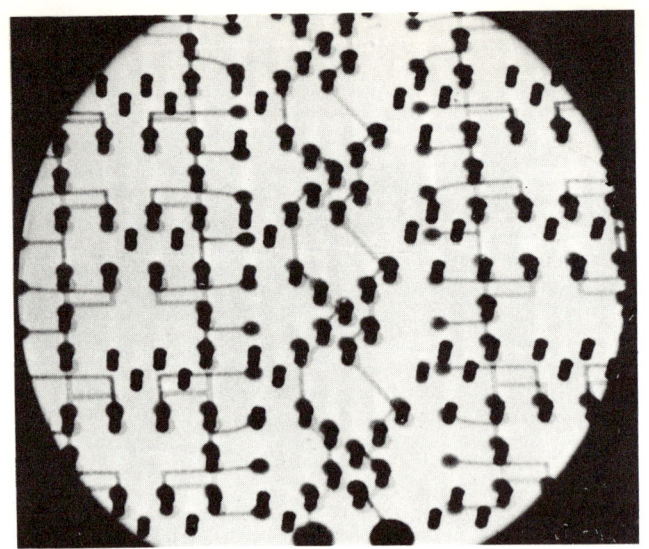

Fig. 13. X-ray top view of pattern.

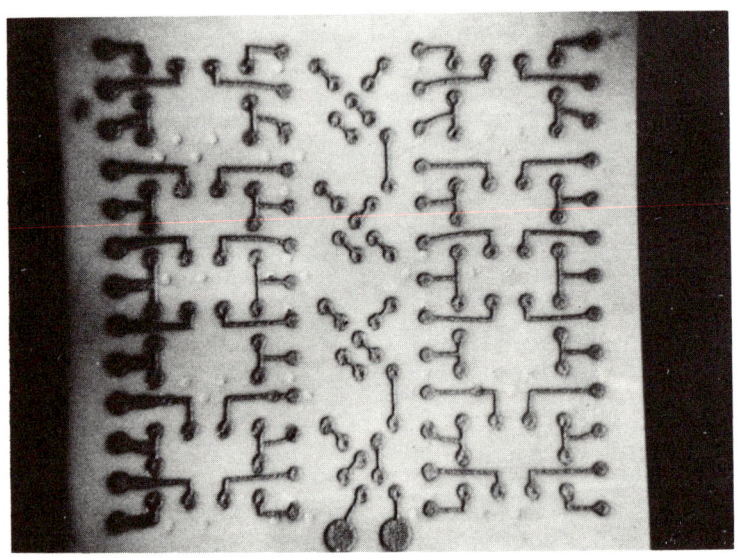

Fig. 14. Top of tinned Ag:Pd vias (no lands).

For the normal 40 mil deep specimens, this provides 12.8 mΩ for each via, or 26 mΩ for both vias. This then permits 30 mΩ for the 40 mil Pd land, or about 6 mΩ/sq. for the approximately 5 squares if we ignore contact resistances. This correlates fairly well with the expected 6-8 mΩ/sq. for internal Pd lands screened in this manner.

Another estimation can be made from two more samples prepared with Pd paste screened into the vias instead of powder, and co-fired with the ceramic. These normal depth (40 mil) holes averaged 40 mΩ. The difference between these and the Ag:Pd powder filled holes is 16 mΩ or 8 mΩ per via. If the 30 mΩ for the Pd lands is subtracted from the 40 mΩ, we get a difference of 5 mΩ per via for the Pd paste filling. If this were bulk Pd, we would expect about 1 mΩ for the 40 mil length for a 14 mil diameter hole. Considering the contact resistances and voided areas in the paste, these numbers correlate well.

If the Ag:Pd powder gave bulk metal resistance, we would expect about 2 mΩ resistance for the same size hole. As mentioned above, we actually obtained 12.8 mΩ or 6 times the bulk value for tinned vias. The measured apparent density of the premalloy 800 is 91.5 gms/cubic inch, or 5.5 gms/cc -- which is about half the theoretical density of Ag:Pd. Sintered land systems commonly provide 2-3 times bulk resistance. Thus, the 6 times bulk density for the powder filling is what might be predicted.

If we assign values of 6 mΩ/sq. for the Pd metallizing and 15 mΩ/sq. for the top Ag:Pd, we calculate resistance of the tested format to be 54 mΩ for each hole with 20 mil long top and internal land sections. We actually obtained 52, 51, 48 and 47 mΩ averages, which is another good correlation.

An approximate 10 mΩ decrease in resistance for this format was observed when the Ag:Pd powder was sintered at 1000°C instead of 900°C. These vias probably were about 7 mΩ instead of the normal 12.8 mΩ and thus only about 4 times bulk resistance.

An analysis of via resistance measurements shows that the via and contact resistances are small when compared to the total land resistance. They should not be of serious concern from this standpoint, but only as to whether the resistances imply a reliability exposure when the modules are stressed.

Conclusion. A relatively simple process has been described which offers considerable economy and ease of implementation over other methods. Powder filling can be used for interconnecting the layers of multilevel ceramics, for interconnecting two sides of a relatively thick ceramic substrate, or for filling transverse core windings. It avoids the complex rheological control often required in paste or liquid systems, and is extremely flexible with respect to format. However, three limitations apply which should be taken into consideration: all vias should be open at the top surface, the holes should be at least about 8 to 10 mils minimum diameter, and the holes should be closed at the bottom to prevent the powder from falling through. The process should be studied further from the aspects of process optimization and reliability. However, in its present state, apparently reliable packages can be readily produced.

BIBLIOGRAPHY

1. M. N. Haller and C. J. Owen, U. S. 3,079,282.
2. W. H. Kohl, "Ceramics and Ceramic-to-Metal Sealing", Vacuum, 14, pp. 333-354, Pergamon Press.
3. R. T. Hopper, "How to Apply Noble Metals to Ceramics", Cer. Ind., June, 1963.
4. L. C. Hoffman, U.S. Patent 3,385,799.
5. I. Haller, IBM, private communication.
6. L. K. Keys et al, "Photoetching and Screen Printing of Conductor Patterns for Face-Down Bonded Devices", Proc. IEEE Electronic Components Conference, May 1970, pp. 87-91.
7. B. Schwartz and D. L. Wilcox, "Laminated Ceramics", Proc. IEEE Electronic Components Conference, May, 1967.
8. J. J. Suran, "A Perspective on Integrated Circuits", IEEE Spectrum, January, 1970, pp. 67-79.
9. R. N. Wild, "Effects of Gold on Solder's Properties", Proc. NEPCON, 1968.
10. O. A. Short, "Conductor Compositions for Fine-Line Screening", Elec. Packaging and Prod., February, 1968, pp. TF9-TF14.
11. D. T. DeCoursey, "Materials for Thick Film Technology - State of the Art", Solid State Technology, June 1968, pp. 29-34.
12. R. E. Thun, "Thick Films or Thin?", IEEE Spectrum, October 1969, pp. 73-79.
13. J. A. O'Connell, "Thick Film Technology", Elec. Pack. and Prod., October 1967, pp. 10-16; December 1967, pp. 13-17.
14. L. F. Miller, U.S. Patent 3,374,110.
15. A. J. Deyrup (Bi_2O_3, Lead Borosilicate Flux), U. S. Patent 2,207,723.
16. J. J. Knox (Bi_2O_3 and Lead Borosilicate), U. S. Patent 2,385,580.
17. G. H. Knox, "Auto Voltage Regulator Typical of New Uses for Thick Film Circuits", Assembly Eng., July 1970, p.35.
18. C. A. Krier and R. H. Jaffe, "Oxidation of the Platinum-Group Metals", J. Less-Common Metals, 5, No. 5, p. 411.
19. B. R. Coles, "The Lattice Spacings of Nickel-Copper and Palladium-Silver Alloys", J. Inst. Metals, 84, 1955-1956, p. 346.

20. Hafner and Volterra, "Palladium:Silver Alloys", Metals Handbook, 8th Edition, Vol. 1, p. 1194.
21. W. H. Aarts and A. S. Houston - MacMillan, "Anomalous Behavior of Ag:Pd Alloys on Plastic Deformation", Acta. Met., 5, Sept. 1957, p. 525.
22. C. N. Rao and K. Krishna Rao, "Effect of Temperature on the Lattice Parameters of some Silver:Palladium Alloys", Can. J. Phys., 42, July, 1964, p. 1336.
23. R. L. Moss and D. H. Thomas, "Formation Structure and Catalytic Activity of Evaporated Palladium:Silver Films", Trans. Far. Soc. 60, No. 498, 1965, pp. 1110-1121.
24. J. N. Pratt, "The Thermodynamic Properties of Silver:Palladium Alloys", Trans. Far. Soc., 56, No. 7, 1960, pp. 975-987.
25. N. H. Nachtrieb, et al., "Self-Diffusion of Silver in Silver:Palladium Alloys", J. Chem. Phys., 26, No. 1, January, 1957, 1. 106.
26. R. L. Rowland and N. H. Nachtrieb, "Self-Diffusion of Palladium in Silver: Palladium Alloys", Trans. Far. Soc., Vol. 67, Dec. 1963.
27. G. H. Laurie and J. N. Pratt, "Electronic Constitution and Partial Thermodynamic Properties of Liquid Tin + Palladium + Silver Alloys".
28. Progress Reports Nos. 32, 40, 42 — National Printing Ink Research Institute, Lehigh Univ.
29. E. M. Davis et al., U.S. Patent 3,292,240.
30. L. C. Hoffman, et al., "Adhesion of Platinum Gold Glaze Conductors", IEEE Comp. Conf. 1965; W. Crossland and L. Hailes, "Thick Film Conductor Adhesion Reliability", 1970 Hybrid Microelectronics Symposium, p. 3.3.
31. Coles and Taylor, "Electrical Resistivity of Ag:Pd Alloys", Proc. Roy. Soc., Series A, 267, p. 139.
32. A. H. Mones and K. Neisser, U. S. 3,248,345.
33. M. L. Block and A. H. Mones, "Properties of Indium Oxide Glaze Resistors", Proc. 1966 IEEE Electronics Comp. Conf., pp. 191-195.
34. I. P. Driear, - "Observations on Formation of Cleavage at Interface Between Glaze Resistor and Termination", IEEE Comp. Conf., 1965.

35. T. F. Eagan and A. Mendizza, "Creeping Silver Sulphide", J. Electrochem. Soc., 101, No. 4, April 1960, p. 353.

36. O. A. Short, "Silver Migration in Electric Circuits", Tele-tech and Electronic Ind., Feb. 1956, p. 64.

37. G. T. Kohman, et al., "Silver Migration in Electrical Insulation", Bell System Tech. J., 34, No. 6, November, 1955, p. 115.

38. E. L. Milne and P. Gibbs, "Ag Ion Migration in Quartz", J. Appl. Phys., 35, No. 8, August, 1964, p.2364.

39. S. W. Chaiken, et al., "Silver Migration and Printed Wiring", Ind. and Eng. Chem., 51, No. 3, March, 1959, p. 299.

40. A. Hornung, "Diffusion of Silver in Borosilicate Glass", Electronic Components Conference, May 8-10, 1968.

41. S. Milkovich and L. F. Miller, U.S. Patent 3,537,892.

42. L. F. Miller, G. J. DePaolo, U.S. Patent 3,414,417.

43. W. Guertler, "A Compendium of Constitutional Ternary Diagrams of the Metallic Systems", WADC Tech. Report 58-615, part 1, March 1959.

44. O. A. Novikova and A. A. Rudnitzkiy, "A Study of the System Gold-Silver-Palladium", Zhur. Neorg. Dhimi, 3, (3), 729-749 (1958).

45. D. L. Herbst, "Composition of Thick Film Resistors", Proc. Int. Soc. Hybrid Microelectronics, 1968, pp. 173-182.

46. F. M. Collins and C. F. Parks, "Thallium Oxide Glaze Resistors", 1967 IEEE Electronics Comp. Conf., pp. 432-438.

47. E. H. Melan and A. H. Mones, "The Glaze Resistor - its Structure and Reliability", 1964, IEEE Electronics Comp. Conf.

48. J. Van Hise, "Process Variables in Thick Film Resistor Fabrication", Proc. Int. Soc. Hybrid Microelectronics, September 1969.

49. L. C. Hoffman, "Precision Glaze Resistors", Ceramic Bulletin, Vol. 42, No. 9, 1963, pp. 490-493.

50. L. Hunt et al., "Sorption of Water Vapor by Hydrophilic Polymers", Official Digest, February 1963, pp. 113-128.

51. A. C. Zettlemoyer, "The Pigment-Vehicle Interface", Official Digest, December 1957, pp. 1239-1271.

52. L. F. Miller and K. N. Neisser, Jr., U. S. Patent 3,390,104.

53. F. M. Clark, "Insulating Materials for Design and Engineering Practice", Wiley, New York, 1962, pp. 414-415.

54. A. K. Doolittle, "The Technology of Solvents and Plasticizers", 1954, Wiley, New York; T. R. Stull, "Vapor Pressure of Pure Substances - Organic Compounds", Industrial and Engineering Chemistry, Vol. 39, #4 April 1947, pp. 517-539.

55. L. F. Miller, U.S. Patent 3,414,641.

56. H. K. Dicken, "Surveying Chip Interconnection Techniques", Elec. Pack. and Prod., Oct. 1970, pp 34-45.

57. L. Curran, "In Search of a Lasting Bond", Electronics, 26, 72-80 (1968).

58. R. K. Field, "The New World of 'Leaded' Chips", Electronic Engineer, 27, 100 (1968).

59. P. R. Clark and H. Baker, "A Diffusion Bonding Program", Technical Report RADC TR 67-62 (April 1967) AD651545.

60. J. S. Cubert et al., "Face Down Bonding of Monolithic Integrated Circuit Logic Arrays", Proc. IEEE Electronic Components Conference (1966), pp. 156-166.

61. E. M. Davis et al., "Solid Logic Technology: Versatile, High Performance Microelectronics", IBM J. Res. Develop., $\underline{8}$, 102, (1964).

62. L. F. Miller, U.S. Patent 3,429,040; U.S. Patent 3,459,133.

63. R. P. Sopher and P. A. Totta, IBM J. Res. Develop., 13, 226 (1969).

64. C. Karan et al., USP 3,401,055.

65. L. S. Goldmann, IBM Components Division, private communication.

66. L. F. Miller, "Microelectronic Device Standoffs", IBM Technical Disclosure Bulletin 8, No. 3 (Aug. 1965), p. 380.

67. K. C. Norris and A. H. Landzberg, IBM J. Res. Develop. 13, 266 (1969).

68. W. Schuelke, "Modular Approach to System Design", Automation 14, 77 (1967).

69. J. Bagrowski et al., "Interconnection of Monolithic Integrated Circuits Through the Use of Advanced Materials and Techniques", IEEE Trans. Parts, Materials and Packaging PMP-2, 90 (1966).

70. E. F. Platz, "Reliability of Hybrid Microelectronics", WESCON Reliability Symposium (IEEE G-PMP), August 21-22, 1968, (Symposium Record, p. 512).

71. R. R. Weirather and C. Go Tiang, "Dielectric Bath Promotes Togetherness in ICs", Electronics 40, 123 (1967).
72. L. F. Miller and R. K. Spielberger, USP 3,401,126.
73. J. A. Baldrey and O. A. Short, USP 2,961,416.
74. A. Wager, IBM Components Division, private communication.
75. J. E. McCormick, "On The Reliability of Microconnections", Elec. Pack. and Prod., June 1968, pp. 187-190.
76. J. Bunton, "The Broken Promise of L.S.I. Packaging", Electronics, March 30, 1970, pp 123-131.
77. I. M. Hymes et al., USP 3,303,393.
78. Article, Electronics, January 4, 1971, "Fairchild Joins the Automated Gang Bonders", p 21.
79. M. P. Lepselter, "Beam-Lead Sealed-Junction Technology", Bell Laboratories Record, Vol. 44, No. 9, Oct.-Nov. 1966, p. 298.
80. Special Issue on Beam-Leads, Western Electric: The Engineer, December 1967.
81. Article, Electronics Review, January 22, 1968.
82. C. Chiou and J. R. Garcia, USP 3,325,882.
83. K. C. Hu, "Application of Polyimide Film in Chip Packaging and Interconnection, 1969 NEPCON Proceedings, 2/1.
84. J. P. Stelmak, USP 3,387,365.
85. C. G. Thornton, "New Trends in Microelectronic Fabrication Technology 1965-1966", SCP and Solid State Technology, March 1966, pp. 42-49.
86. A. G. Cozens and J. E. Tomlin, "High Density Soldered Interconnections", Microelectronics, July 1968, p. 38.
87. M. P. Eleftherion et al., "Handling and Bonding of Beam Lead Sealed Junction Integrated Circuits", Proc. IEEE Components Conf., 1968, p. 149.
88. Article, "Tabs Ease Bonding", Electronics, March 4, 1965, pp. 190-191.
89. H. K. Dicken and D. B. Kret, "Assembling Integrated Circuits", Electronic Engineer, October 1966, pp. 66-71.
90. E. F. Koshinz, "Semiconductor: Wire Bonding and Face Bonding Considerations and Comparisons", Proc. SAE Microelectronic Packaging Conf., Nov. 1968.

91. H. S. Evander, "Flip Chips - A Practical Scheme for LSI", American Ceramic Society, April 1968 Presentation.
92. Article, Electronic News, October 23, 1967.
93. R. P. Moore, USP 3,391,451.
94. P. J. A. McKeown, and R. D. Peacock, "Flip-Chip Semiconductor Devices", Electrical Communication, 41, No. 4, 1966, p. 431.
95. W. B. Hugle et al, "Flip Chip Assembly", Solid State Technology, Aug. 1969, pp 62-67, 99.
96. C. P. Sandbank and P. J. A. McKeown, "New Interconnection Techniques for Multichip and Hybrid Integrated Circuits", Proc. IEEE, Dec. 1964, pp. 1655-1657.
97. L. D. Bernstein, "Brazing, Soldering and Joining Materials by the Solid-Liquid Interdiffusion (SLID) Process", Extended Abstract 133, Electrochem. Soc. Proc., May 1965.
98. J. Napier and R. F. Ross, Jr., USP 3,392,442.
99. M. Weissenstern et al., USP 3,392,442.
100. G. W. V. Lane, "Materials for Conductive Elements", IEEE Proceedings, 1967, Paper 37.4, p. 306.
101. A. R. Riben, "Microbonds for Hybrid Microcircuits", AD647464, January 1967.
102. D. L. Hirsch, "The Percussive Arc Welder", Elec. Pack. and Prod., May, 1968, pp. 206-209.
103. F. Z. Keister et al., "Interconnection Techniques for Microcircuits", IEEE Trans. on Component Parts, March 1964, pp. 34-41.
104. J. G. Bouchard and M. A. Ponti, Sr., "Flip-Chip Bonding and Dimensional Transformation - A Low Cost Microcircuit Production Technique", Proc., Hybrid Microelectronics Symp., June 1967, B 3.
105. T. J. Matcovich, "Direct Bonding Processes and Hybrid Circuit Applications", Elec. Pack. and Prod., Nov. 1967, p. 101.
106. R. P. Moore, "An Evaluation Study of Ultrasonic Face Bonding of Integrated Circuit Chips to Evaporated Aluminum Conductors", Proc. IEEE Components Conference, 1968, pp. 427-435.
107. C. Chiou et al., "Decal Interconnection Bonding Method", IBM TDB, Vol. 10, No. 7, December 1967, p. 1074.

108. M. H. Schwab and C. Wurms, "Multichip Bonding", IBM TDB, Vol. 10, No. 8, January 1968, p. 1247.
109. G. K. Fehr, "Microcircuit Packaging and Hermetic Sealing", 1968 WESCON Proceedings 6/2.
110. G. R. Giedd and A. F. Karsch, "Indium Bond for Silicon Chip Attachment", IBM TDB, Vol. 11, No. 2, July 1968, p. 117.
111. E. Q. Carr, USP 3,289,046.
112. W. K. Spielmann and L. Rauch, "Repair Technology for Modules", IBM TDB, Vol. 10, No. 11, April 1968, p. 1810.
113. F. P. Heiman, "Integrated Logic Nets", AD 646726, Jan. 1967.
114. W. B. Archey, "Hot Gas Soldering for Interconnection of Integrated Circuit Packages", Proc. IEEE, December 1964, p. 1657.
115. Article, "Getting the Right Bump on the Right Pad", Electronics, Feb. 19, 1968.
116. Article, Electronics, February 19, 1968.
117. M. E. Szekely, USP 3,114,867.
118. P. Scharf et al, "Flip-Component Technology", Proc. 1967 IEEE Components Conference, pp. 269-275.
119. S. Oktay, "Parametic Study of Temperature Profiles in Chips Joined by Controlled Chip Collapse Techniques", IBM J. Res. Develop., Vol. 13, May 1969.
120. J. H. Butler et al., USP 3,365,620.
121. W. E. Harding et al., USP 3, 280,019.
122. S. Wagner and M. Walker, "Low Cost Integrated Circuit Techniques", AD 635183, July 1966.
123. S. Wagner and W. Doelp, "Low Cost Integrated Circuit Techniques", AD 652699, May 1967.
124. L. F. Miller, "Elongated Flexible Chip Joint", IBM TDB, Vol. 10, No. 11, April 1968, p. 1610.
125. C. Chiou and F. L. Graner, "Making Decal Interconnections to Semiconductors and Substrates", IBM TDB, Vol. 8, No. 11, April 1966, p. 1541.
126. C. R. Cooke et al., "For Low Cost Flat Packs: Remove the Wires, Put Circuits in Glass", Electronics, July 12, 1965, pp. 99-104.

127. J. Marley and G. Tolson, "New Beam-Lead Connection Method Boosts Semiconductor Memory Yields", Electronics, Dec. 22, 1969, pp 105-110.
128. Article, "Facing up to the Chip", Electronics Review, July 1967.
129. Article, Electronics News, February 19, 1968.
130. Article, A. Socolovski et al., Electronic Engineer, April 1968, pp. 92-95.
131. P. Mallery, "Thermal Pulse Bonding of Beam Leads", Proc. IEEE Components Conference, 1967, pp. 276-282.
132. F. L. Howland, "Bonding Beam-Lead Silicon Integrated Circuits", 1968 NEPCON Proceedings.
133. H. M. Wagner, "Welding Fine Leads to Thin Films", Elec. Pack. and Prod., March 1966, pp. 20-28.
134. A. P. Broyer, "Flex-Cable Interconnections Mass Bonded with Infrared", Proc. NEPCON 1967, pp. 29-32.
135. H. D. Doolittle et al., "Ceramic-Metallizing Tape for Reliable Metal-Ceramic Seals", IEEE Trans. on Electronic Devices, September 1961, pp. 390-393.
136. J. H. McCusker, USP 3,331,125.
137. W. L. Doelp et al., USP 3,374,537.
138. D. W. Crawford et al., "Localized Microwave Heated Dielectric Glass", IBM TDB, Vol. 10, No. 3, Aug. 1967, p. 325.
139. G. P. Lang, USP 3,316,628.
140. B. G. Bender et al., USP 3,316,628.
141. C. R. Tichner, "Chip Removal by Hot Gas", IBM TDB, Vol. 11, No. 7, Dec. 1968, p. 875.
142. L. F. Miller and J. M. Schiller, "Semiconductor Chip Joining", IBM TDB, Vol. 10, No. 5, October 1967, p. 657.
143. R. Helda, "Spider Bonding Technique with I.C.s", Proc. Int. Electronic Circuit Pack. Symp. WESCON, Aug. 20-21, 1969, 4/3.
144. A. P. Mandel and P. K. Vahey, "Thermal Design Criteria for Hybrid Microelectronic Modules", Proc. Hybrid Microelectronics Symp., Oct. 28-30, 1968, p. 31-42.
145. See, for example, D. R. Fewer, Proc. IEEE, Electronic Comp. Conf., 1966; T. R. Meyers, "Flip-Chip Microcircuit Bonding Systems", Proc. IEEE Comp. Conf., April 30-May 2, 1969, pp. 131-144.

146. A. Coucoulas and B. H. Cranston, "Compliant Bonding-A New Technique for Forming Microelectronic Components", IEEE Trans. Electronic Dev. DE-15, No. 9, Sept. 1968, p. 664-674.

147. I.C.E., Inc., "Integrated Circuit Engineering-Basic Techniques", Section 7.15, 1966 Ed.

148. R. H. Cushman, "Mechanical Thermal Pulse Metal Joining", Symposium Record, 7th International Electronics Circuit Packaging Symposium, 1966, p. 4-12.

149. J. E. Clark, "Wobble Table for Thermocompression Bonding Beam-Lead Silicon Integrated Circuits", Proc. 1968, Int. Electronic Circuit Pack. Symp., Aug. 19-20, 1968, p. 211.

150. D. S. Peck, "Reliability of Beam-Lead Sealed-Junction Devices", Proc. 1969 Ann. Symposium Reliability. IEEE # 69C8, pp 31-44b.

151. R. Naylor, "Study of Flip-Chip and Beam Lead Microcircuit Assemblies", Proc. IEE Vol. 116, 1969.

152. D. Boswell, "Mechanical Design of Chip Components for 'Flip' and Short Beam Lead Mounting", Proc. Hybrid Microelectronics Symp., Sept. 29-30, 1969, pp. 5-11.

153. B. Dale, "The Application of Beam-Lead Devices to Hybrid Microcircuits", Proc. 1968 IEEE Elec. Comp. Conf., pp. 84-92.

154. M. B. Shamash, "New Device Bonding Techniques," Presentation NEPCON, Feb. 12, 1969.

155. G. E. Gorgenyi, "The Hughes Dimensionally Controlled Glass Ambient Flip Chips," presented at NEPCON, Feb. 12, 1969.

156. P. Lin et al., "Design Considerations of a Flip-Chip Joining Technique", Proc. Hybrid Microelectronics Symp., Sept. 29-30 1959, pp. 313-322.

157. L. F. Miller and A. H. Mones, "Chip Capacitor Configuration", IBM Technical Disclosure Bulletin Vol. 10, No. 7, December 1967, p. 941.

158. C. M. McIntosh, U.S. Patent 3,540,894.

Original Citations

Chapter 1 — "Paste Transfer in the Screening Process", Proc. S.A.E. Microelectronic Packaging Conference, November 20-22, 1968, pp 26-33.

Chapter 2 — "Survey of Thick Film Electrodes", 70th American Ceramic Society Meeting, April 1968.

Chapter 3 — "Silver Palladium Fired Electrodes", Proc. 1968 IEEE Electronics Components Conference, pp 52-64.

Chapter 4 — "Ternary Alloy Electrode Pastes", Proc. 1968 WESCON Electronic Circuit Packaging Symposium, Section 3/2.

Chapter 5 — "Glaze Resistor Paste Preparation", Proc. 1970 IEEE Electronic Components Conference, pp 92-101.

Chapter 6 — "Controlled Collapse Reflow Chip Joining", IBM J. Development, Vol. 13, No. 9, May 1969, pp 239-250;

"Joining Semiconductor Devices with Ductile Pads", 1968 Proc. International Society for Hybrid Microelectronics, pp 333-342;

"Module Materials for Controlled Collapse Chip Joining", Proc. 1969 NEPCON, pp 432-445.

Chapter 7 — "A Survey of Chip Joining Techniques", Proc. 1969 IEEE Electronic Components Conference, pp 60-76.

Chapter 8 — "A Critique of Chip Joining Techniques", Solid State Technology, Vol. 13, No. 4, April 1970, pp 50-62.

Chpater 9 — "Powder Interconnections", 1969 Proc. WESCON Electronic Circuit Packaging Synposium, Section 2/2.

INDEX

Adhesion of conductors, 25, 28, 39, 45, 47, 61, 63, 74, 75, 78, 115-117, 121, 125
    Cohesion, 35, 36, 44, 116
Agglomeration of particles, 34, 39, 62, 81, 94, 98
Air cooling, 177
Air isolation, 181
Alignment of chips, 110, 175
Alloying, 25, 50, 122
Ammonium sulphate, 37, 43, 47, 64
Apparent density of powders, 34
Automation, 165, 174

Back-joined chips, Chaps. 7 and 8
Ball bonds, 137
Barium acetate, 120
Barium flouride-borate flux, 120
Beam leads, 130, 134, 147, 153, Chaps. 7 and 8
Beam substrates, 150
Beilby layer, 119
Bentonite, 86
Binder (see also resin), 7, 83, 85
Bismuth oxide, 35, 39, 45, 53, 62, 63, 65, 66, 68
Bond strength, 170
Borosilicate glass, 35, 45, 54, 62, 63, 68, 120
Bumps for devices, 134, 137
Butt connections, powder, 184
Butyl carbitol acetate, 37, 75, 87, 93

Capacitive effects of chip pads, 108
Capacitors, 27, 53
    Chip capacitors, 54
Capillary infiltration, conductors, 25, 27, 187
    By glass, 36
Carbonaceous residues, 87, 90, 93
Casson plots, 75
Chip handling, 165, 177
Chip joining, Chaps. 6-8
Chromium-copper-gold, 102, 139
Ceramics: multilevel, 27, Chap. 9
    Glazed substrates, 47, 116, 118, 121
    Surfaces, 36

Cermet, 25
Cohesive strength, 35, 36, 44, 116
Cofiring, 192
Colloidal silica, 36, 118, 119
Colloidal powders, 118
    effect on rheology, 121
Compressive strength of conductors, 44
Conductance, 33, 38, 39, 41, 78, 114, 120, 122, 131
Conductors, 131, 189, Chaps. 2, 3, 4
Connection pins, 33
Contact resistance, 44, 66, 68
Controlled collapse chip joining, Chaps. 6-8
Copper conductors, 27, 28, 52, 58
Copper balls, 130, 134
Copper conductivity, 111, 124
Corrosion, 37, 56
Cost, 27, 38, 56, 62, 70, 74, 77, 142, 158, 167, 168
Crossovers, 170
Cured conductors, 25, 29

Dams as solder stopoffs, 104, 106, 121-123
Decals, 130, 132, 151
Decomposing salts, 118
Density of conductors (see also porosity, fissuring), 40, 41, 61, 70, 73, 74
Density of metals, 120
Design flexibility (chip joining), 160
Dicing of wafers, 176
Dielectrics, 27
Diffusion chip joining, 136
Dilution of metals in conductors, 120
Dispersion of powders, 39, 47, 51, 62, 64, 75, 93-99
Dots, 104-106, 123-126
Drift of resistors, 53, 83
Drying of pastes, 43, 51, 64, 78, 81
Ductility of joints, 101, 112

Edge connections, powder, 184
Edge shorting of devices, 101, 106
Electron beam, 137
Electron microprobe, 36, 116, 119
Epoxy, 26, 139
Erosion by solder, 34, 38, 50, 55, 61, 62, 63, 69, 74, 124
Etching of conductors, 27
Ethyl cellulose, 37, 65, 75, 84, 87
Eutectic bonds, 138, 139, 140, 151
Extra pads for chip joining, 104, 108
Expansion coefficient, 36

Filling the moat, 151
Firing, 25, 27, 36, 43, 70, 75, 78, 81, 92, 93, 98, 117, 192

　　　　In inert atmosphere, 117, 192
　　　　In reducing atmosphere, 27, 28, 117
　　　　Dams, 122
　　　　Cofiring, 192
　　　　Refiring, 75, 77
Fissuring of conductors, 37, 43, 62, 64, 74, 68, 76, 77
Flexible joints, 173
Flip chips, 101, 128, Chaps. 6-8
Flow control of pastes, 7, 36, 39, 47
Flux (also frit), 25, 28, 33, 35, 39, 45, 53, 54, 63, 68, 69, 75, 76, 116, 119-120
Formulation, Ag/Pd paste, 33
Frequency dependence of resistors, 95
Furoic acid, 19, 37, 43, 47
Furnaces, 27

Gelation agents, 7, 36, 37, 44, 63, 74, 75, 86
Germanium/silicon, 139
Glass: crossovers, 54, 78
　　　(see also dams)
　　　Coloration, 45
　　　Viscosity, 36
　　　Pastes, 122, 139
　　　Interaction with conductors, 55
Glazed ceramic, 116, 118, 121
Gold: conductors, 28, 73, 119, 124, Chap. 4
　　　/platinum palladium, Chap. 4
　　　/palladium, Chap. 2
　　　/platinum, 11, 28, 36, 40, 43, 49, 51, 56, 61, 116, 119, Chap. 4
　　　/powders, 63, 190
　　　/silicon, 139, 140, 151
Grind gage, 96

Heat dissipation, 113, 121, 130, 142, 168
Heat sinks, 113, 130, 142, 179
Heating techniques, 140
Hermetic packages, 148
Humidity, 86
Hydrogenated castor oil, 75, 86

Imbedded devices, Chap. 8
Immersion cooling, 113, 142, 178
Indium, 139
Infiltration, 25, 27, 187
Inductance problems, 182
Infrared heating, 140
Ink oil, 84
Inside connections on chips, 169, 179
Inspectability of joints, 171
Intermetallic compounds, 25, 28, 50, 52, 57, 117
Isolated lands for chip joining, 108

Joint strength, 111, 112, 137, 138, 170

Laminated chip pads, 147
Lasers, 137
Lead borosilicate glass, 35, 45, 54, 62, 63, 68, 120
Lead frames, 130, 133-134, 148-150
Line width limitations, 141
Liquid cooling, 113, 142, 178

Magnesium, 58, 117
Malleable pads, 130
Masks, 5, 114
    Meshless, 5, 15-17, 19, 22
    Etched mesh, 15-17, 22
    Thickness, 15
Migration, silver, 38, 53, 58, 68, 117, 124, 125
Milling pastes, 39, 47, 51, 63, 75, 77, 93-99
Mixing pastes, 39, 96
Molybdenum, 27
Moly/manganese, 25, 27, 28, 121
Morphology of conductors, 47, 61, 81
Multilevel ceramics, 27, 183, Chap. 9

Nickel, plated, 27
Nitrogen curtain, 146
Non-tinnable paste, 105, 117-121
Non-tinning agents, 118-119
Nonyl phenoxy polyoxyethylene ethanol, 37

Oil absorption value, 118
Optical characteristics - resistors, 97
Organometallic compounds, 25, 118
Overlap method, 104, 106
Oxides, 118, 123
    On surfaces, 34, 38
    In glasses, 35

Palladium, 13, 14, 15, 20, 23, 27, 34, 73, 85, 190
    Oxide, 34, 123
Particle size, 33, 39, 63, 74, 81, 96, 189-190
Paste immobilization in screening, 13, 14, 15, 23
    In other printing processes, 12
Peritectic joining, 136
Petroleum distillates, 87
Photosensitive binder, 25, 27
Planarity of contacts, 101, 110, 137, 147, 150
Platinum, 27, 28, 65, 119
Polyalphamethyl styrene, 87, 93
Polyisobutylene, 90
Polyimides, 25, 123
Polymethylmethacrylate, 87

Polystyrene, 87, 93
Polyterpenes, 87
Porosity of conductors, 40, 69
Potassium, 83
Powders, 25, 189-192
    Alloyed, 25, 39, 191
    Gold, 63, 190
    Heat treating, 62
    Palladium, 34, 76
    Platinum, 62, 74
    Settling, 43, 64, 70
    Silver, 34, 189
Powder interconnections, Chap. 9
Power dissipation, 177-178
Pre-alloyed powders, 25, 39, 190
Printing processes, 12, 141
Process criticality-chip joining, 176
Process control, 163, 171
Profile of screened deposits, 44, 48

Rapid pulse reflow, 146
Real estate on wafers, 159, 181
Reduction-silver, by hydrazine, 34
Reflow process, 111
Refrigeration of paste, 47
Registration, 119, 125, 133, 165
Reliability, 57, 58, 112, 127, 158, 161-163, 166
Repair of conductors, 139-140, 162, 193
Replacement of chips, 137, 139-141, 168
Resinates, 25
Resins, 81, 85, 87, 93, Chap. 5
Resistance of conductors (see also conductance), 29, 48-50, 51, 115, 120
    Of powder filled holes, 199
Resistors, 27, 65, Chap. 5
    Control, Chap. 5
    Doped, 43, 66, 73, 83, 96, 98
    Effect of electrode, 52, 53, 81
    Effect of vehicle, 91-94
    Indium oxide, 43, 81, 98
    Oxidation, 93-94
    Silver/palladium oxide, 66, 68, 81
Rheology, 5, 7, 13, 15, 17-19, 23, 34, 44, 74, 81, 84, 90
    Changes in, 47
    Effect of particle size, 34
    Plug flow, 19
    Secondary flow, 57, 75
    Shear rates, 13
    Stability, 39, 47, 74
    Thixotropy, 18, 19, 44
    Viscosity, 17, 18, 86, 89
    Yield value, 7, 18, 44, 75

Scaling of resistors, 52, 53, 66
    Effect of electrodes, 52, 66
Screening, Chap. 1, 5, 34
    Clogging, 17, 34, 39, 49
    Depression of screens, 14
    Emulsions, 11, 83
    Masks, 5, 15-17, 19, 22, 114
    Mesh, 5, 8, 11
    Process, 22, 81
    Squeegee, 14, 22
Self-alignment of chips, 110, 175
Silicon real estate, 159, 181
Silicone, decomposed, 120
Silver, 28, 34, 68
    /gold:palladium, Chap. 4
    /migration, 38, 53, 58, 68, 117, 124, 125
    /nitrate, 34
    /palladium, 25, 28, 77, 116, 119, 124, 191, Chap. 3
    /platinum, 25, 119
    /sulfide, 56
    /bearing solder, 125
Sintering, 25, 40, 43, 49, 65
SLT, 101, 111, 130, 134, 144
Sodium, 83
Sodium chloride, 37
Solder reflow, 101, Chap. 6
    Of device pads, 102, 135, 144
Solids loading, 63, 81
Solvents, 83, 87
Spacing, chip to substrate, 112
Spider bonding, Chap. 8
Standoffs on chips, 146
STD process, 152, Chap. 8
Stringing of paste, 85
Squeegee, 5, 12, 13, 14, 22, 44
Solder/flux, 63
    /lead tin, 55
    /processes, 40, 55
    /silver/lead/tin, 38, 47
    /thicknesses, 51, 56
    /wetting, 56
Subliming solids, 37
Surface area, 34, 73
Surface tension, 108, 119

Tapered pads, 137, 175
TCR, 50, 53, 93, 95
Temperature heirarchies, 136, 171
Tensile strength, 28, 36, 45, 115-117
Terephthallic acid, 19, 37, 43, 47, 64
Terpineol, 84, 187

Thermal expansion, 112
Thermogravimetric analysis, 37, 65
Thermocompression bonding, 101, 128, 134, 137, 140, 146, 151
Thin films, 132, 180
Tin, 38, 41
Tinnability, 29, 34, 38, 47, 55, 69, 70, 74, 77, 123, 125
Titanium, 25
Transfer, Chap. 1, 5, 8
Transverse channels, 184
Tungsten, 27, 28, 121

Ultrasonic bonding, 101, 130, 134, 138, 147

Vacuum metallizing, 135, 136, 151, 152
Vehicle, 7, 28, 33, 36, 64, 83
    Controlled flow. 20
    Drying, 25
    Hydrophilicity, 86
    Immobilization pigment, 34
    Polar, 7, 83, 86
    Pyrolysis, 36, 37, 65, 93
    Reactivity, 85
Vibration soldering, 55
Viscosity (see rheology), 17, 18, 86
Viscosity coefficient of temperature, 89
Volatility, 36, 64, 75, 83, 87
Volume of paste ingredients, 77

Wedge bonds, 137
Welding, 137
Wetting agents, 34, 37, 64, 83
Wires, 134, 151
Wire bonding, Chaps. 7, 8

Yields, chip joining, 159
    Powder filling, 187, 196

X-ray, 38, 62, 77

Zirconium, 25